WORKBOOK

Endorsed for learner support

Cambridge
International AS & A Level

Chemistry
Practical Skills

Judy Potter
Olli Dunning

HODDER
EDUCATION

The Publishers would like to thank the following for permission to reproduce copyright material.

Photo credits: Hazard symbols on p5, p53 and p54 © jusep/stock.adobe.com

Acknowledgements

Every effort has been made to trace all copyright holders, but if any have been inadvertently overlooked, the Publishers will be pleased to make the necessary arrangements at the first opportunity.

The information in this book is based on the Cambridge International AS & A Level Chemistry (9701) syllabus for examination from 2022. You should always refer to the appropriate syllabus document for the year of your/students' examination to confirm the details and for more information. The syllabus document is available on the Cambridge International website at www.cambridgeinternational.org.

Orders: please contact Bookpoint Ltd, 130 Park Drive, Milton Park, Abingdon, Oxon OX14 4SE. Telephone: +44 (0)1235 827827. Fax: +44 (0)1235 400401. Email education@bookpoint.co.uk Lines are open from 9 a.m. to 5 p.m., Monday to Saturday, with a 24-hour message answering service. You can also order through our website: www.hoddereducation.co.uk

ISBN: 9781510482852

First published in 2020 by
Hodder Education,
An Hachette UK Company
Carmelite House
50 Victoria Embankment
London EC4Y 0DZ

www.hoddereducation.co.uk

Impression number 10 9 8 7 6 5 4 3 2 1

Year 2024 2023 2022 2021 2020

Cover photo © School of Chemistry, University of Edinburgh

Illustrations by Aptara Inc.

Typeset in India by Aptara Inc.

Printed in the UK

A catalogue record for this title is available from the British Library.

Contents

Introduction

The excitement of chemistry starts with the practical work. To make sense of the trends in properties of chemicals and to be able to predict or explain a reaction, careful and detailed observations need to be made. These observations may be qualitative, such as a colour change, bubbles being produced or a precipitate being formed; they may be quantitative such as the measurement of mass, volume or a temperature change.

This workbook focuses on providing you with the practical skills and knowledge required by the Cambridge International AS & A Level Chemistry syllabus (9701). Following a chapter on safety, the book is divided into two halves: Chapters 2 to 4 cover the skills and knowledge required by the AS Level qualification; Chapters 5 to 7 cover the additional skills and knowledge you will require if you are sitting the full A Level. Each of these two halves is further divided into the following chapters:

- **Skills**: these chapters cover some of the key skills required by the syllabus, including taking measurements, recording results and drawing conclusions.
- **Methods**: these chapters discuss some of the practical investigations that you are likely to encounter in your studies.
- **Practice questions**: these are provided to give you an opportunity to check your understanding and put the skills you have learned into practice. Some are taken from previous Cambridge International AS & A Level Chemistry examination papers; others have been written in the style of examination questions. As with an exam paper, there are spaces for you to write your calculations and answers in the book. Answers to these questions can be found online at **hoddereducation.com/cambridgeextras**.

1 Safety

Before carrying out any practical work, the safety aspects need to be considered; although your teachers will have carried out a risk assessment for all experiments, it is important that you too know the risks involved in a practical and how to minimise them.

- Never take food or drink into the laboratory, tie back long hair, do not apply make-up, smoke or deal with a contact lens. If you have been handling chemicals or apparatus, keep your hands away from your face.
- Know and follow the rules of the laboratory. Note where safety equipment such as fire blankets, extinguishers and eye wash stations are situated and make sure you know who to contact in case of an emergency.
- Wear safety spectacles and a laboratory coat. Keep your own laboratory space neat as you are less likely to spill solutions or mix the wrong reagents.
- Read and make sure you understand your instructions. Some reagents which you use may have hazard labels attached. Note these and follow suitable precautions when handling such materials:

Figure 1.1 Hazard labels (from left to right): corrosive, environmental hazard, flammable, oxidising, toxic

- Ensure you are mixing the correct reagents and carefully follow instructions.
- Heating with a Bunsen burner carries the risk of burns. Boiling liquid too quickly may result in hot liquid splashing out of a vessel such as a test-tube. To minimise this risk, move the test-tube in the flame to ensure more even heating and make sure the mouth of the test-tube points away from yourself – and from anyone else. Check to see if a boiling tube has been provided – these are wide test-tubes – as boiling liquids are much less likely to boil over or spit from these.
- If any chemical enters your eyes, immediately wash the affected eye well at an eye wash station or sink and inform a member of staff.
- Report any accident, such as breakage of glassware or spillage of chemicals, to staff. Dispose of any broken glass in the appropriate bins to minimise the risk of anyone getting cut by it.
- At the end of practical work, leave your working area clean and tidy. Wash your hands well before leaving the laboratory.

AS LEVEL

2 Skills

Units of measurement

The most common units used in chemistry are presented in Table 2.1.

Measured quantity	Name of unit	Symbol
Length	metre	m
	centimetre	cm
Mass	gram	g
	kilogram	kg
	tonne	t
Volume	cubic metres	m^3
	cubic centimetres	cm^3
	cubic decimetres	dm^3
Concentration	gram per cubic centimetre	$g\,cm^{-3}$
	gram per cubic decimetre	$g\,dm^{-3}$
	moles per cubic decimetre	$mol\,dm^{-3}$
Time	second	s
Energy	joule	J
	kilojoule	kJ
Pressure	pascal	Pa
	kilopascal	kPa
	atmospheres	atm
Temperature	Kelvin	K
	Celsius	°C

Table 2.1 Common units of measurement

Using stock solutions

Stock solutions are often provided in laboratories. These are solutions that have been accurately made up to particular concentrations. You may take solution from the stock solutions as directed, but any unused excess should be discarded and never returned to the stock solution bottle as this may contaminate the solution. Do not draw into a pipette from a stock solution; always pour it into a clean beaker and draw from there.

Using apparatus

Ensure you have practised with and know how to use and read the various implements for measurement, and that you are familiar with the procedures they will be used for.

Thermometer

The **thermometer** in use in Figure 2.1 has gradations every 1 °C. When reading it, ensure the thermometer is vertical and your eye is horizontal with the reading.

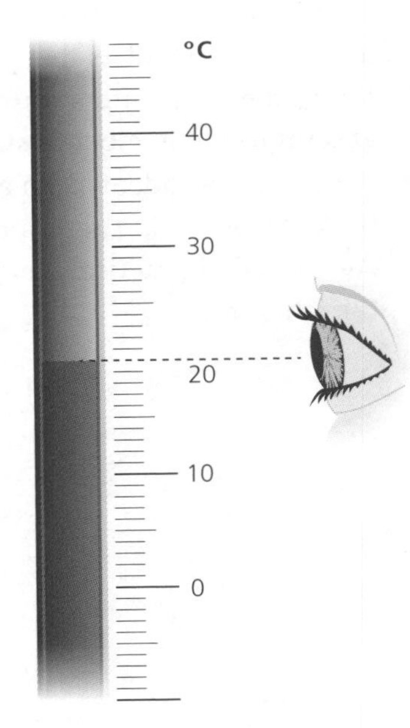

Figure 2.1 How to read a thermometer

Using this thermometer, it is possible to judge the reading to the nearest 0.5 °C (half a gradation mark) and the temperature should be recorded to this precision.

To show the accuracy of the reading, the figure after the decimal point should be given, even if it is zero (0); a temperature could be recorded using this thermometer as 20.0 °C, but **not** 20 °C or 20.00 °C.

Top-pan balance

A **top-pan balance** can normally weigh up to 200 g to the nearest 0.01 g and two significant figures after the decimal point are required when recording a mass. Make sure the pan is clean and that the reading starts at 0.00 g. Weigh into a small beaker, weighing boat or weighing bottle to avoid contamination. Use a clean spatula.

For many experimental methods, masses, such as the mass of chemical added, need to be measured accurately.

If a specific mass is required for the experiment, adding it is straightforward:

1. Put a container onto a top-pan balance and zero the balance. This sets the mass reading to zero.
2. Add the chemical using a spatula (if solid) until the mass required is shown.
3. Transfer the chemical to the experiment.

If the mass to be used is not specific but still needs to be measured accurately, the method used is **measuring by difference**.

1. Weigh the mass of the chemical and its container.
2. Add the chemical to the experiment.
3. Weigh the mass of the (now empty) container.

The difference between the two masses is the mass of chemical added to the experiment. This takes into account any chemical remaining in the container. If you need to measure the mass of a solid or liquid accurately, this is the method to use.

Stop-clock

For timing, a **clock** or **watch** needs to be used, normally read to the nearest second. If you are to use a stopwatch, ensure you know how to use it before you start timing any practical experiment. Occasionally, you may be asked to measure time in minutes, so make sure you know which it is to be before you start timing.

Pipette

The 25 cm³ and 10 cm³ **pipettes** are precision glassware used for measuring these two volumes accurately to within one drop, which is 0.05 cm³. When filling a pipette, always use a pipette filler; these come in various types, so ensure you know how to use the type which is issued to you.

The pipette should be cleaned by rinsing with a little deionised water and then rinsed with a little of the solution to be used, which is then discarded. The solution is then drawn into the pipette using a pipette filler until the meniscus is above the gradation mark (Figure 2.2). Remove the filler, keep the pipette vertical and your line of sight horizontal with the gradation and, using a finger to control the flow, allow solution to drip slowly out of the pipette until the bottom of the meniscus is on the gradation.

Allow the measured solution to run out of the pipette into the reaction vessel and when no more flows, dip the pipette end underneath the liquid surface, which draws a little more from the pipette to make up the correct volume. The last remaining drop should not be shaken or blown out of the pipette as the tiny amount of liquid left in the pipette is allowed for in the calibration.

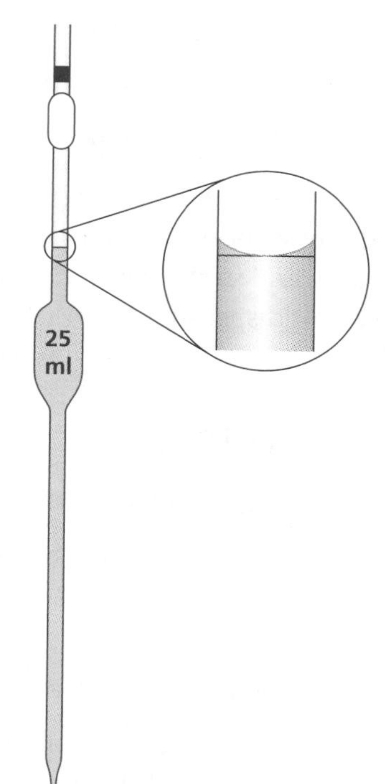

Figure 2.2 How to fill a pipette

Burette

A **burette** is also a piece of precision glassware which allows a volume up to 50 cm^3 to be measured. The burette should be cleaned by rinsing it with a little deionised water, allowing the water to run out past the tap. It should then be rinsed with a small amount of the solution which is to be used in it and this should again be allowed to flow out through the tap and be discarded. Ensuring the tap is closed, the burette is clamped vertically and filled to above the gradations with the help of a filter funnel. The filter funnel is removed so that it cannot drip into the burette and alter the readings. Solution is run out using the tap so that the meniscus is on the scale and the portion of tube below the tap is also filled with solution. It is important to ensure there are no air bubbles in the solution below the tap. If there are, they can be removed by gently shaking the burette vertically over a sink while the tap is open.

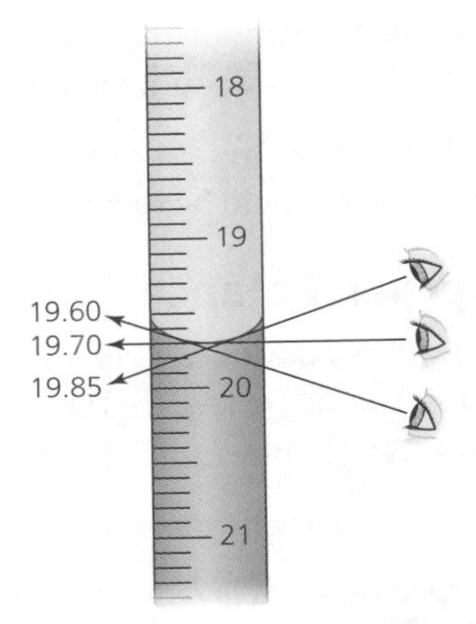

Figure 2.3 How to read a burette

Note that the gradations start at 0.0 cm^3 at the top and are marked in 1.0 cm^3 intervals to 50.0 cm^3 at the bottom. The intervals are further divided into 0.1 cm^3. The eye has to be horizontal across the gradation or error is introduced as shown in Figure 2.3. Readings on the gradation marks are taken to 0.00 cm^3 and those in between are taken to 0.05 cm^3. Looking horizontally across the meniscus and gradations, the reading is taken at the bottom of the meniscus to 0.05 cm^3.

The only exception to taking the reading at the bottom of the meniscus is when potassium manganate(VII) is used in the burette. The strong purple colour can mask the reading and in this case all readings are taken from the top of the meniscus.

Volumetric flask

A **volumetric flask** is a glass (or plastic) vessel calibrated to contain a precise volume of solution at a particular temperature; flasks come in various sizes, the most common ones that you will come across being 50 cm^3 and 250 cm^3 although there are smaller and larger ones than these. The volume of the flask is shown on the side and the flask has a gradation line on its neck (Figure 2.4); the accuracy is also printed on the side of the vessel. A flask filled so that the bottom of the meniscus is on that gradation line contains the precise volume of solution. A tightly fitting stopper is needed.

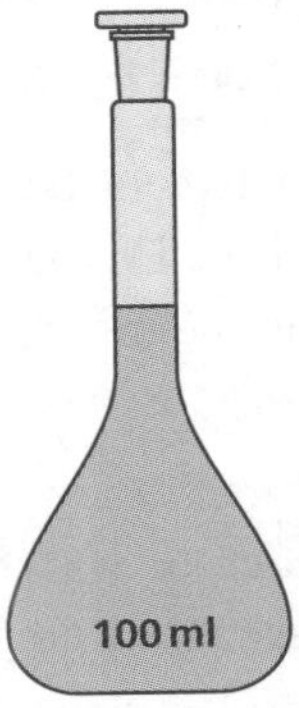

Figure 2.4 A volumetric flask

A volumetric flask is used to prepare a **standard solution** – one for which the concentration is accurately known.

Measuring cylinder

If the high precision of a titration is not required in the volume measurement, a **measuring cylinder** may be used. These measure a volume of solution with less precision and are often used when the reagent is to be added in excess and so the exact volume is not required. These come in various sizes; normally you will use those capable of dispensing 10 cm^3, 20 cm^3, 25 cm^3 or 50 cm^3. Gradation marks indicate volumes 1 cm^3 apart and values in between them can be estimated. At best, the accuracy is to one decimal place.

Procedures

Making up a standard solution

When finding a concentration, you normally need to make up a standard solution first. This is a solution of known concentration that you use in titrations against a solution of unknown concentration.

1 Determine the volume of solution you will need. For titrations, it is normal to make 250 cm^3 of solution to allow for several titration measurements.

2 Calculate the mass of pure solid chemical you need to dissolve into the volume in step 1 for the required concentration. Take into account waters of crystallisation (e.g. 0.0100 mole of $CuSO_4.5H_2O$ is 2.495 g, not 1.595 g).

3 Place a clean, dry beaker (normally 50 cm^3) onto a top-pan balance.
4 Add the required mass of pure solid to the beaker using a clean spatula and then remove from the balance.
5 Add approximately 40 cm^3 of distilled water and stir with a glass rod until fully dissolved.
6 Transfer the solution to the volumetric flask using a glass funnel.
7 Rinse the beaker and glass rod with more distilled water and transfer the washings to the volumetric flask.
8 Remove the funnel and add distilled water directly to the volumetric flask quickly until the liquid is just starting to rise into the flask neck. Insert the stopper and invert several times to ensure the contents are well mixed. Then, with the flask on a horizontal surface, carry on adding water drop by drop until the bottom of the meniscus rests on the gradation mark (this is called 'making up to the mark').
9 Stopper and invert the volumetric flask several times to fully mix the solution.
10 Label the flask.

This method ensures all of the solid is transferred into the volumetric flask to give a solution of accurately known concentration.

Once the standard solution is made, accurate dilutions can be prepared by pipetting out a known volume into a second volumetric flask and making up the volume with deionised water.

Heating solids or solutions

When heating a test-tube or boiling tube containing liquid, use a test-tube holder and keep the tube moving in the flame to try to avoid sudden boiling and ejection of liquid. A heavy precipitate may also cause sudden boiling and care needs to be taken as the temperature is raised. Heating a small amount of solid needs to be approached in a similar manner, by keeping the tube moving in the flame. If it is suspected that steam or water is a product, clamping the test-tube horizontally so the clamp is near the mouth of the test-tube and then holding the Bunsen to move the flame over the part containing the solid will prevent moisture condensing on cold parts of the tube, running back into the hot part and cracking the tube.

Adding reagents

For observational exercises, use appropriate amounts of chemicals – for solids this is usually a small spatulaful and for solutions 1 to 2 cm depth in a test-tube is generally sufficient. Keep your spatula clean and dry to prevent cross contamination. Add solutions using a teat pipette, which should also be clean but not necessarily dry. Keep the teat pipette in a small beaker of distilled water to be ready for use and wash it after each use so that there is no cross contamination. Practise using it with the thumb and first finger so that the barrel lies across your other fingers and control the addition of solutions drop by drop.

Mix chemicals by gentle shaking – and **never** by placing a thumb over the top of the tube.

Testing for gases

There are five gases you may be asked to identify – hydrogen, carbon dioxide, oxygen, chlorine and ammonia.

If hydrogen is suspected, as it is so much lighter than air, a boiling tube or test-tube held with a test-tube holder over the source of hydrogen can be used to collect enough gas to test; the test-tube can then be inverted near a Bunsen to provide a good squeaky pop.

If carbon dioxide is suspected, the gas can be bubbled through lime water which turns milky (Figure 2.5). A quick way of testing is to use a dry teat pipette to collect a portion of gas in the reaction tube and then bubble this captured gas through a small amount of lime water in a test-tube.

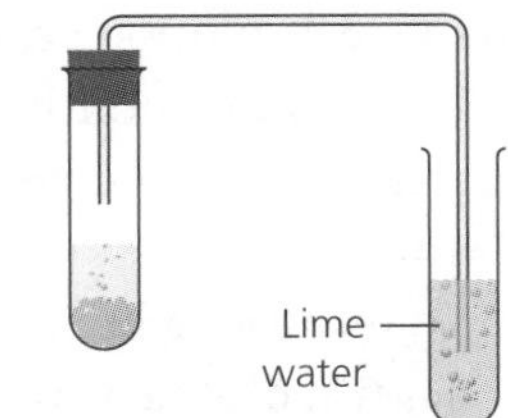

Figure 2.5 Testing for carbon dioxide gas

If oxygen gas is suspected as a product, a wooden splint lit and then extinguished so that it is just glowing, inserted into the test-tube, will reignite.

Chlorine bleaches moist litmus paper placed at the mouth of the test-tube.

The last of these, ammonia, will make its presence felt by its pungent odour and can be tested by a piece of damp red litmus paper placed at the mouth of the test-tube where the litmus will turn blue.

Filtration

Leave a hot solution to cool to allow as much solid as possible to separate out. If solid fails to separate, scratching the inside of the test-tube with a glass rod may be required to start crystallisation. The filter paper should be folded into four and then opened to fit into the filter funnel (Figure 2.6).

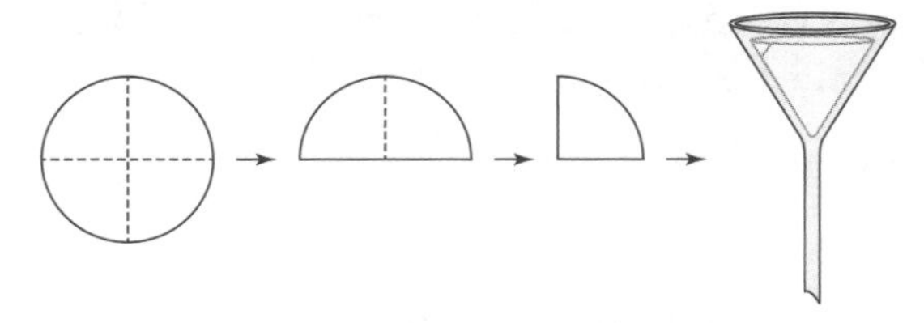

Figure 2.6 How to fold a filter paper for filtration

A Büchner funnel uses air pressure to push the solution through the filter paper and can allow the process to take place more quickly.

The Büchner funnel has a flat bottom pierced with holes and a special filter paper is used to just cover this. The filter paper is placed in position and then moistened with a few drops of the solvent used. The funnel is then placed on the flask – the seal should make an airtight join. The flask is then connected to an air or water pump and the lower pressure inside the flask allows the air pressure outside to push the solvent through the paper. Gently shake the mixture to be filtered to bring crystals into the body of the solution and then pour it down a glass rod onto the centre of the filter paper. It may be necessary to return some filtrate to the test-tube to wash out the remaining crystals or simply scrape any remaining crystals onto the filter paper. Allow air to be drawn through the filter paper and crystals for some minutes to dry them.

After filtration, remove the filter paper and crystals and allow them to air dry, or dry them in an oven or desiccator. Then gently remove the crystals into a clean container but do not scratch the filter paper as this contaminates the crystals with paper fibres.

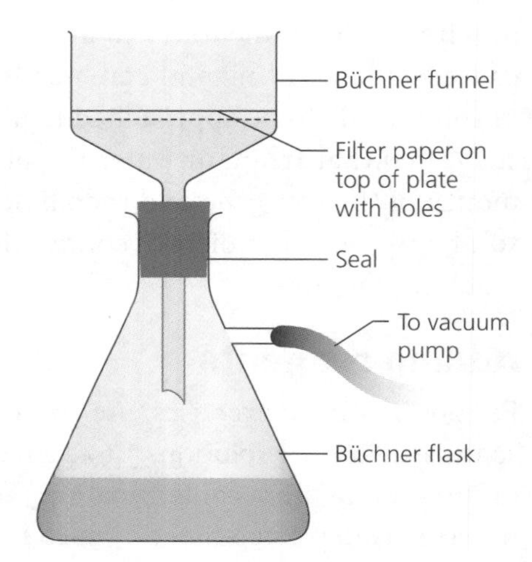

Figure 2.7 Büchner funnel

Testing for ions in solution

The main tests are for acid–base nature, tests to indicate which anions are present, tests to identify which cations are present, and the tests for oxidising and reducing agents.

- Start with about 1 to 2 cm depth of solution in a test-tube. The acidity can be determined with a piece of moist litmus paper.
- The carbonate, halides, sulfate, sulfite and thiosulfate ions can be determined by adding a few drops of the test solution according to Table 8.2 (tests on anions, page 93) using a teat pipette.
- For nitrate and nitrite, to the solution in a boiling tube, add dilute NaOH until the tube is about a quarter full and add a small spatulaful of aluminium powder. Warm the mixture gently and test for ammonia.
- To test for reducing agents such as nitrite and iodide, to 1 cm depth of solution in a test-tube, add acidified potassium manganate(VII) dropwise using a teat pipette to see if the purple reagent is decolourised.

- To test for oxidising agents, 1 cm depth of solution is placed in a test-tube and aqueous potassium iodide solution is added dropwise using a teat pipette. The dark brown colour of iodine is seen, which can be tested by the formation of the dark blue colour of the complex formed when fresh starch solution is added to the mixture.
- Testing for the cations requires addition of sodium hydroxide solution and ammonia solution to separate samples of the solution. To a 1 to 2 cm depth of solution in a test-tube is added NaOH(aq) dropwise with a teat pipette until there is no further change. Note any precipitate formed and, on adding excess NaOH(aq) until the tube is about half full of solution, note if the precipitate dissolves. Warm the mixture gently but take care if there is a precipitate as bumping may occur.
- The test with NH_3(aq) is carried out in a similar manner but with **no** warming as this would drive ammonia gas out of the solution.

Recording quantitative data and observations

It is important to have a suitable space in your notes – maybe a table ready drawn up – so that data and observations can be recorded at the time or as soon after the observation as possible. It is difficult to remember the exact colour of a precipitate or how a solution changed by the time you have come to the end of a piece of practical work.

All numerical data collected, including every burette reading, should be shown in a suitable form, which may be in a table of results.

Every piece of numerical data should be identified with correct units and label or description, which is often in the form of a heading in a table.

Presentation of data or observations

A **graph** may be the appropriate way to present the data. A well-drawn graph is very accurate.

- It is important to use sensible scales. A typical piece of A4 graph paper has 9 × 7 large squares. Look at the largest values to be plotted on the two axes and select an appropriate scale for each. For example, if the largest values are 175 and 60, use 9 and 6 large squares; the 9 will be labelled to 180 (20 units per large square) and the 6 labelled to 60 (10 units per large square). A large square is easily divided into 2, 5 or 10; avoid divisions into 3 or 4 which are difficult to deal with. The scales need to be linear – check that equal unit divisions are the same number of squares.
- Draw the two axes with a sharp pencil and put in the main divisions with a short line and a number. If the origin is included put zeros (0) on both axes.
- Label the axes with the name of what is being plotted and the units; use a solidus (/) to separate the units from the name.
- One variable is **independent** of the experiment being performed. In chemistry this is often time or volume added, which is plotted on the *x* axis (the horizontal one). The axis would be labelled 'time/s' or 'volume/cm^3'. The other variable is **dependent** on the experiment being performed and is plotted on the *y* axis (the vertical one). Likely dependent variables are temperature (from a cooling curve), concentration or rate (from a kinetics experiment) and pH (from a titration curve).
- Put the points to be plotted in a table at the top right corner of the graph if there is room – or in a clear table near to the graph. The first column is usually the dependent variable (*y* axis), the second the independent variable (*x* axis). The table needs to have headings with the appropriate units. Make sure that the scale you have chosen will allow all points to be plotted, but also allow for any extrapolations to continue and finish on the graph paper.
- Plot the points in pencil (use a sharp HB or F lead), using a small circled dot. (In the unlikely event of having to plot two graphs on the same sheet of paper, use small crosses to indicate the second set of points.) If you need to include the origin, make sure it is there. By inspection, decide if the points lie on a straight line or a curve. If the points lie on a straight line, use a ruler for the best fit; for a curve you will probably have to draw freehand.

You may be asked to obtain data from the graph, such as the following:

1 Read off a flat portion (e.g. from a cooling curve).
2 Work out a gradient (e.g. from a concentration against time graph in kinetics to obtain the rate). This is easy if the graph is a straight line; it is much harder if it is a curve as you will have to draw a tangent to the graph at the point selected. This is very difficult to get right and the result is more unreliable.
3 Read off the value of an intermediate point (**interpolation**, which is reliable) or of a point found by extending the graph (**extrapolation**, which is less reliable).
4 Make sure you give the result to an appropriate degree of accuracy.

Distinguish between a graph and a diagram. A plot of melting point of elements against proton number is not a graph. The points may be joined up to show a trend but they do not lie on a line of best fit as there are no 'intermediate elements'. However, diagrams can be very helpful in illustrating trends or patterns in behaviour of the elements and compounds.

Performing calculations

Every stage in a calculation should be clearly set out.

Calculated values should be shown to an appropriate number of significant figures – an answer cannot be more accurate or shown to a greater degree of accuracy than the data from which it is obtained. In general, the answer needs to be to the same number of significant figures as the data; one more significant figure should be carried through the calculations. If the data being used is known to 3 or 4 significant figures, then the final calculated value will be known to 3 significant figures, but the calculation will be carried out to 4 or more significant figures.

All calculated values should have the appropriate units.

When carrying out multiplication and division with values that have been measured, the answer has the same number of significant figures as the value with least significant figures.

EXAMPLE 2.1

Finding the number of moles in a portion of a solution:

Titre = 16.35 cm^3

Concentration of solution = 0.11 mol dm^{-3}

Number of moles = (0.11/1000) × 16.35 = 0.0017985 on your calculator

But the concentration value has the lesser significant figures (2) and so the answer should have 2 significant figures. By rounding appropriately, the number of moles = 0.0018.

EXAMPLE 2.2

Finding a percentage purity when experimental work leads to a mass of 2.3789 g on your calculator, but this is the mass contained in 2.5 g of impure substance.

The percentage purity = (2.3789/2.5) × 100% = 95.156% on your calculator

But the mass of the impure substance is only known to 2 significant figures and so your final answer should be given to 2 significant figures, which is 95%.

When adding and subtracting, your answer should have the same precision as the least precise measurement used.

EXAMPLE 2.3

Adding two volumes together 6.38 cm^3 + 8.2 cm^3 = 14.58 cm^3 on your calculator

But one of the measurements has 2 significant figures and the other has 3 significant figures and so your answer should be given to 2 significant figures, rounded appropriately, 14.6 cm^3.

EXAMPLE 2.4

Subtracting initial from final readings of a volume from a burette

Final reading/cm^3	25.65	25.65	25.60	5.65	25.0
Initial reading/cm^3	3.0	3.00	3.60	5.05	13.00
Calculated volume run in/cm^3	22.65	22.65	22	0.6	12
Corrected volume run in/cm^3	22.7	22.65	22.00	0.60	12.0

Identifying sources of error and suggesting improvements

You need to be able to distinguish between systematic errors and random errors.

Systematic errors may be introduced by the method itself, or by the use of a piece of equipment:

- Sources of error introduced by the method itself need to be identified and you should suggest how the method could be modified to minimise such errors. For example, carrying out only one titration and therefore producing only one titre value would not allow a demonstration of reproducibility and so further titrations would be suggested in order to improve on this.
- Using different balances for a series of weight measurements could introduce a zero error. Using the same balance for all such measurements would eliminate this source of error.

Random errors should also be identified. For example, a change of temperature during a rates experiment will alter the rate.

A statement of 'human errors' is not acceptable; though there are occasionally errors arising in the observer's ability to observe, e.g. in the disappearing cross experiment, which would be a random error.

For the purpose of the Cambridge International AS & A level Chemistry syllabus, the maximum uncertainty in a quantitative measurement is half the difference between the closest calibrations, e.g. for a thermometer calibrated at °C the maximum uncertainty is ±0.5 °C, therefore the maximum percentage error in a temperature change of 14.0 °C = ((2 × 0.5)/14.0) × 100 = 7.14%.

Uncertainties in measurements should be estimated:

- A two-decimal-place balance measures to the nearest 0.01 g. The uncertainty in the measurement is half way between two readings = 0.005 g.
- For a measurement of 15.00 g, the uncertainty = (0.005/15.00) × 100% = 0.033%.
- For a measurement of 1.50 g, the uncertainty = 0.33%.
- For a measurement of 0.15 g, the uncertainty = 3.3%.

AS LEVEL

3 Methods

Titration

Titration is a method of finding an unknown mass or concentration using accurately measured volumes of solutions that react together. The volume of one solution is accurately measured using a burette as it is added to another solution, the volume of which has been accurately measured by pipette into a reaction vessel. There is normally a visual method of noting when the reaction is complete, but in some titrations the pH or temperature change may be measured instead. Usually in titrations, we work with concentrations between about 0.5 to 0.05 mol dm^{-3} and often need to adjust a concentration by dilution to give a suitable titre value.

The reaction vessel, usually a conical flask, needs to be clean but does not need to be dry.

Use a pipette to transfer a fixed volume of solution into the conical flask and place it on a white tile under the burette. Add a few drops of indicator.

Record the initial burette reading and add the solution of known concentration from the burette to the conical flask carefully while swirling the mixture until the colour of the indicator changes. Near the end-point, add from the burette one drop at a time – you should be able to see the colour change on the addition of just one drop of solution.

Record the final burette reading. The difference between the two readings is known as the 'titre'.

Repeat steps 3–6 until three concordant titres (titres within 0.10 cm^3 of each other) are obtained. The average of the concordant titres is the volume of known concentration solution needed to react fully with the unknown concentration solution.

The choice of indicator required depends on the chemical reaction between the two solutions. It is important that the indicator has an abrupt colour change at the equivalence point of the reaction. (The equivalence point is when the reaction is complete; the end-point is when the colour change occurs and is a property of the indicator.)

If the concentrations of the two solutions are very different, for example in a titration of 0.10 mol dm^{-3} NaOH with 1.5 mol dm^{-3} HCl, 10.0 cm^3 NaOH solution pipetted would require only 0.70 cm^3 HCl solution to reach the equivalence point. The smaller the value of the titre, the greater the percentage error and so it is preferable to accurately dilute the HCl solution to achieve a greater titre value in the titration.

To dilute 1.5 mol dm^{-3} HCl to 0.15 mol dm^{-3} HCl, 10.0 cm^3 of the 1.5 mol dm^{-3} HCl is pipetted into a clean 100 cm^3 volumetric flask and deionised water is added up to the mark. The flask is then well shaken.

Acid–base titrations

Titrations of acids against bases are the most common and the simplest to understand. One solution neutralises the other and the indicator is chosen with the pH of its end-point to match the pH of the equivalence point.

You need to be able to identify a strong acid and know how the pH changes when a strong acid is titrated with a strong or weak base; similarly you need to be able to identify a weak acid and know how the pH changes when a weak acid is titrated with a strong base. An acid–base indicator should change colour (the end-point) at the equivalence point (when the exact amount of reagents are mixed). In each of the three possible cases, the pH at the equivalence point can be identified:

- a strong acid titrated against a strong base – pH at the equivalence point = 7.0
- a strong acid titrated against a weak base – pH at the equivalence point is below 7.0
- a weak acid titrated against a strong base – pH at the equivalence point is above 7.0.

Table 3.1 lists the common indicators and their suitability for use in acid–base titrations. The indicator range is the pH range over which the indicator changes colour. The end-point is in the middle of this range.

			Indicator range/pH	
Acid	**Base**	**Indicator**	**From**	**To**
strong	strong	methyl orange or bromophenol blue or thymol blue or thymolphthalein	3.1 (red) 3.0 (yellow) 8.0 (yellow) 9.3 (colourless)	4.4 (yellow) 4.5 (blue) 9.6 (blue) 10.5 (blue)
strong	weak	methyl orange or bromophenol blue	3.1 (red) 3.0 (yellow)	4.4 (yellow) 4.5 (blue)
weak	strong	thymol blue thymolphthalein	8.0 (yellow) 9.3 (colourless)	9.6 (blue) 10.5 (blue)
weak	weak	no suitable indicator	n/a	n/a

Table 3.1 Common indicators and their suitability for use in acid–base titrations

Titrations with methyl orange are usually complete when the colour of the solution is orange – midway between the red and yellow.

Titrations between weak acids and weak bases have no suitable indicator as the change in pH is too gradual, meaning any colour change would not be abrupt enough for a reliable volume reading. It is possible to use a pH meter to monitor the pH change in this case, or a thermometric titration could be attempted (see page 16).

In acid–base titrations, the acid is preferably in the burette. As the indicators are themselves weak acids, it is important to add only a very small amount – usually three drops – in titrations.

Rather than using an indicator, the pH of the solution (measured with a pH meter) can be used to monitor acid–base titrations. The pH is a measure of acidity. Strong acids are fully ionised to produce H^+ ions in solution, but weak acids are only slightly ionised and produce far fewer H^+ ions in solution. The greater the concentration of H^+ ions, the lower the pH. A strong acid solution has a lower pH than a weak acid solution of the same concentration.

Similarly, strong bases are fully ionised to produce OH^- ions in solution and a strong base solution has a higher pH than a weak base solution of the same concentration. The H^+ and OH^- in solution are linked so that as one decreases, the other increases and so usually we deal with the concentration of H^+ only, by measuring the pH. At 298 K, the pH of pure water = 7.0; acid solutions have pH less than 7; alkaline solutions have pH greater than 7.

Redox titrations

Redox titrations may not involve an indicator if the manganate(VII) ion, MnO_4^-, is a reagent. This is a common oxidising agent and is a very intense purple in solution. On reacting, it forms the manganese(II) ion, Mn^{2+}, which is a very pale pink colour in solution. If the experiment begins with $KMnO_4$(aq) in the burette, the end-point is when the first permanent pink colour is seen in the conical flask. This occurs when the reaction is complete and is caused by a small excess of unreacted $KMnO_4$(aq). If the experiment begins with the $KMnO_4$(aq) in the conical flask, the end-point occurs when the intense pink colour just disappears.

Iodine titrations

In sodium thiosulfate–iodine titrations, 2 cm^3 of a freshly made starch solution is used as the indicator. This forms an intense blue–black coloured solution with iodine. The titration is carried out with the thiosulfate in the burette until the reaction mixture is the pale straw colour of very dilute iodine. At this point, the starch indicator is added and the titration continued until one drop turns the (now) dark blue–black reaction mixture colourless.

Thermometric titrations

Thermometric titrations require no indicator but the reaction needs to be exothermic enough for a measurable temperature change. The method is similar to that for an acid–base titration, but an insulated (polystyrene) cup is used instead of a conical flask. The reagent is run in from a burette $1\,cm^3$ at a time while the cup is stirred with the thermometer. After addition of each cubic centimetre, the temperature of the reactant solution is measured and recorded.

Initially, addition of reagent from the burette causes a temperature rise as it reacts in the cup. This causes the initial upward-sloping line on a graph of volume added versus temperature for the first additions. Once equivalence has been reached and the reaction is complete, addition of more reagent from the burette causes the temperature in the cup to fall steadily. By extrapolating the first (heating) trend and the second (cooling) trend, the point at which equivalence is reached can be found (see Figure 3.1).

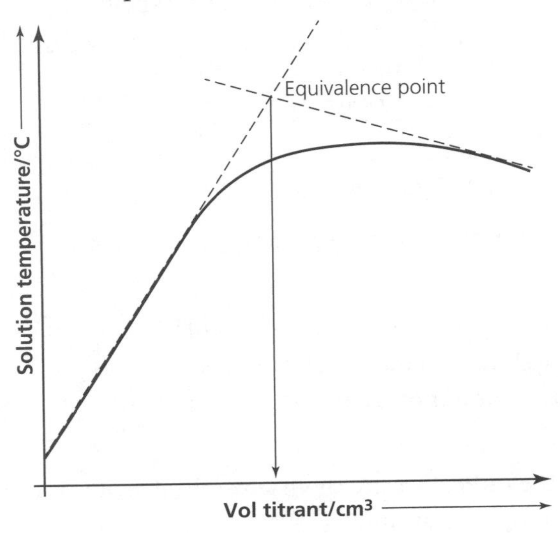

Figure 3.1 Thermometric titration graph

Gravimetric experiments

This type of experiment relies on accurate weighing to determine the answer.

EXAMPLE 3.1

Sodium bromide crystals contain water of crystallisation, $NaBr.nH_2O$. To find n in the formula, a mass of sodium bromide crystals is accurately weighed in a clean beaker. The crystals are then dissolved in water and excess silver nitrate solution is added to precipitate silver bromide, AgBr. The mixture is filtered and the residue washed with a little cold water before being dried in an oven and then reweighed.

- starting mass of $NaBr.nH_2O = 0.791\,g$
- $40\,cm^3$ of distilled water is added and the mixture stirred with a glass rod until the crystals dissolve.
- $20\,cm^3$ of $0.5\,mol\,dm^{-3}$ silver nitrate is added.

The mixture is filtered and the beaker washed into the filter paper with a little distilled water, so that all the precipitate ends up on the filter paper. The filter paper and contents are dried and the crystals are then carefully removed onto a clean, weighed watch glass.

Note that each mass, both those from the balance and the mass of AgBr found by subtraction, is taken to the same number of decimal places. The calculation relies on correct equations and the relationship between the mass, the number of moles and the mass of a mole of each chemical.

mass of watch glass	= 9.736 g
mass of watch glass + AgBr	= 10.787 g
mass of AgBr	= 1.051 g

From the equations,

$$NaBr.nH_2O(s) \rightarrow Na^+(aq) + Br^-(aq) + nH_2O(l)$$

and,

$$Br^-(aq) + Ag^+(aq) \rightarrow AgBr(s)$$

$$\text{number of moles AgBr} = \frac{\text{mass}}{M_r} = \frac{1.051}{187.8}$$

$$= \frac{\text{mass of NaBr.}n\text{H}_2\text{O}}{M_r(\text{NaBr.}n\text{H}_2\text{O})}$$

number of moles AgBr = number of moles Br^- = number of moles $NaBr.nH_2O$

$$\frac{1.051}{187.8} = \left(\frac{0.791}{M_r\,[\text{NaBr.}n\text{H}_2\text{O}]}\right)$$

where, $M_r\,[NaBr.nH_2O] = 141.3$

mass of water in 1 mol of $NaBr.nH_2O = 141.3 - (23.0 + 79.9) = 38.4\,g$

$\frac{38.4}{18} = 2.13$; assuming n is a whole number, $n = 2$

Measuring gas volumes

Experiments to measure gas volumes are used to determine quantities reacting or rates of reaction. In this example, the experiment is to find the volume of 1.0 mol of CO_2. It relies on a correctly balanced equation, accurate weighing and a method of collecting and measuring the volume of the CO_2 gas. The reagents to produce the carbon dioxide are sodium carbonate and hydrochloric acid.

$$Na_2CO_3(s) + 2HCl(aq) \rightarrow 2NaCl(aq) + CO_2(g) + H_2O(l)$$

There are two ways the gas can be collected, as shown in Figure 3.2.

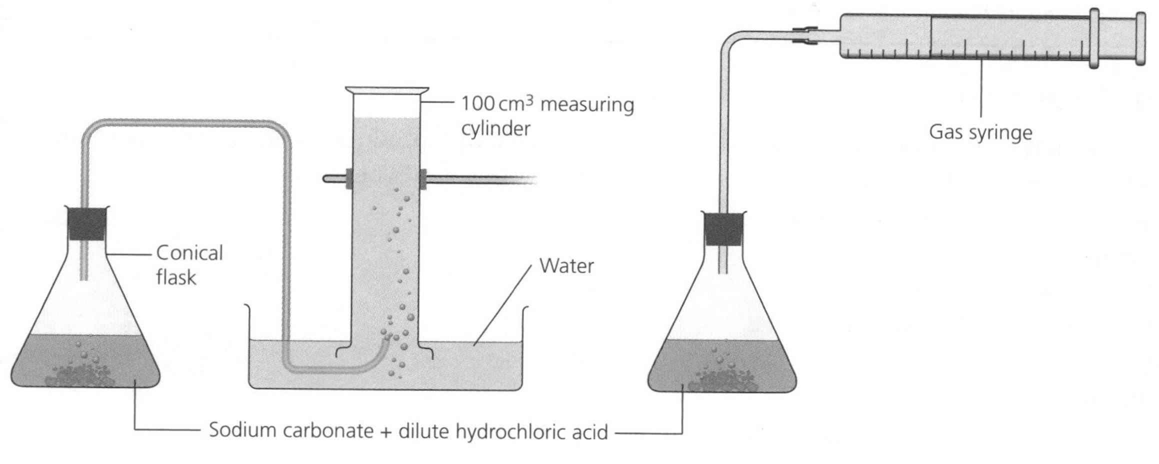

Figure 3.2 Collecting gas

The left-hand apparatus collects the gas over water in an upturned measuring cylinder. The right-hand apparatus allows the gas to be collected without being bubbled through water.

The sodium carbonate needs to be pure – if it is a sample left out in the air, some water will be absorbed from the atmosphere. The hydrochloric acid is in excess.

- Measure 50 cm^3 of 1 $mol\,dm^{-3}$ HCl and pour this into the conical flask.
- Weigh accurately 0.38 g Na_2CO_3 in a weighing bottle or other suitable flask.
- Assemble your apparatus; if using the measuring cylinder, it needs to be full of water and then inverted and clamped over the delivery tube. If using the gas syringe, it needs to be emptied of air. Make sure the apparatus is airtight.
- Take the readings of the measuring cylinder or the gas syringe (often some small amount of air is left in either apparatus and is difficult to remove).
- Add the weighed sodium carbonate and reassemble the apparatus as quickly as possible.
- Reweigh the weighing bottle.
- When the reaction is complete, read the final volume of gas collected in either the measuring cylinder or the gas syringe.
- Calculate the number of moles of Na_2CO_3 used and the volume of CO_2 collected.
- Using the equation, work out the number of moles of CO_2 produced.
- Calculate the volume of 1 mole of CO_2.

EXAMPLE 3.2

Table 3.2 shows the results for two experiments, each using a different method of gas collection.

	Collection of gas over water	**Collection of gas using gas syringe**
volume of 1 $mol\,dm^{-3}$ HCl/cm^3	50	50
mass of weighing bottle/g	8.973	9.024
mass of weighing bottle + Na_2CO_3/g	9.347	9.398
mass of weighing bottle + remaining solid/g	8.976	9.029
mass of Na_2CO_3 used/g	**0.371**	**0.369**
initial gas volume/cm^3	3.5	2.0
final gas volume/cm^3	86.5	86.0
volume of CO_2/cm^3	**83.0**	**84.0**
no of moles Na_2CO_3 = mass/M_r	$(0.371/106) = 3.50 \times 10^{-3}$	$(0.369/106) = 3.48 \times 10^{-3}$
volume of 1.0 mol CO_2/dm^3	$(83.0/3.5 \times 10^{-3})/1000 =$ **23.7**	$(84.0/3.48 \times 10^{-3})/1000 =$ **24.1**

Table 3.2

You would need to suggest an explanation for why the two results are not identical. Possibilities are that the CO_2 was not measured at the same temperature or that some CO_2 dissolved in the water.

Rate of reaction

The rate of reaction is how quickly a reagent is used up or how quickly a product is made. There are many ways of following reactions to find the rate – some of the features that could be used are:

- the disappearance of a coloured reagent
- the appearance of a coloured product
- the volume of a gas that is produced
- the change in volume of gaseous reagents
- the formation of a precipitate
- the change of pH
- the change in mass of reacting mixture as a gas is being given off.

Inspection of the equation for the reaction will give an idea of which method to use.

For example, consider the reaction,

$$CaCO_3(s) + 2HCl(aq) \rightarrow CaCl_2(aq) + H_2O(l) + CO_2(g)$$

The volume of $CO_2(g)$ produced or the loss in mass as the reaction proceeds could be used. These two are more convenient for experimentation than monitoring the pH change which also takes place.

The interesting thing about the rate of a reaction is that it changes as the reaction proceeds; usually the reaction just slows down and then stops, either when all of a reagent is used up or if the reaction reaches an equilibrium position. Some reactions speed up before slowing down as one of the products acts as a catalyst. A change in temperature also changes the reaction rate. To be able to compare reaction rates, just one variable is changed at a time.

For the calcium carbonate reaction, the temperature can be kept constant by immersing the reaction vessel (usually a conical flask or a beaker) in a constant temperature enclosure or at least in a water bath. If the calcium carbonate is used as largish chunks, the surface area (where the reaction occurs) changes very little as the reaction proceeds and so can be taken as constant. Nothing else is added to the mixture, so the only variable is the concentration of hydrochloric acid.

There are two ways of following the reaction rate using the volume of gas produced; both require a stopwatch and the apparatus for gas collection using a gas syringe.

1 **Continuous monitoring**
  - A couple of largish chunks of $CaCO_3$ are added to the conical flask.
  - $50\,cm^3$ of $1.0\,mol\,dm^{-3}$ HCl is measured in a measuring cylinder.
  - The acid is added to the conical flask and the apparatus reassembled quickly. The initial gas volume is read and the stopwatch started.
  - The volume of gas in the syringe is taken at known time intervals until no more CO_2 is being produced.
  - The volume of $CO_2(g)$ for each reading is calculated by taking away the initial gas volume reading.
  - A graph is plotted of the volume of $CO_2(g)$ against the time. This will show an increase up to the maximum (Figure 3.3).
  - The rate of reaction can be found at any time by drawing a tangent to the curve and calculating its gradient.
  - This can then be repeated for other HCl concentrations but keeping the same marble chunks.

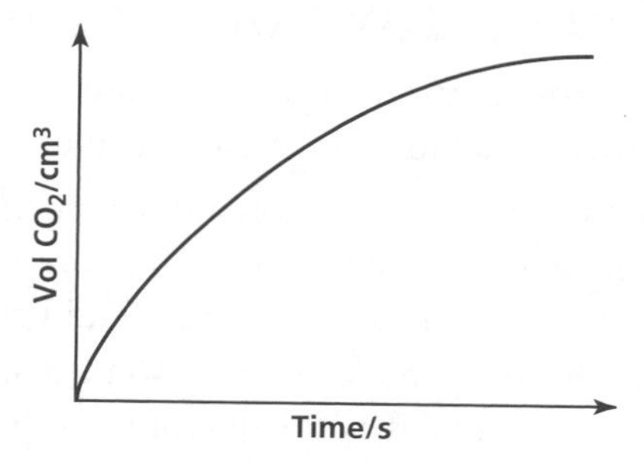

Figure 3.3 Reaction progress graph

2 **Initial rate**

  This method will work as long as the reaction is stopped soon after it has started – it assumes that the HCl concentration does not change over the period of timing. The apparatus is the same, but this time the reaction is timed to a certain point – say until $10\,cm^3$ of CO_2 is produced.
  - Assemble the apparatus with the $CaCO_3$ chunks as before.
  - Measure $50\,cm^3$ of $1.0\,mol\,dm^{-3}$ HCl in a measuring cylinder.
  - Add the acid to the conical flask, reassemble the apparatus, take the original gas syringe reading and start the stopwatch.
  - Take the time, t, when the volume of CO_2 reaches $10\,cm^3$ (allow for the initial gas syringe reading).
  - Dilute the HCl to $0.5\,mol\,dm^{-3}$ and repeat the experiment.
  - Repeat the experiment with three further dilutions.
  - The rate of reaction is proportional to $1/t$.
  - Plot a graph of $1/t$ against the HCl concentration.

Quality of measurements

Titration

It is quicker and leads to greater accuracy to take the initial burette reading when the meniscus is on the scale, rather than trying to get the meniscus exactly on the $0.00\,cm^3$ mark.

All burette readings, both initial and final readings including those from any rough 'range finder' titration, should be recorded to the correct number of significant figures.

Titrations should be carried out until two titres agree; this may not be possible in the time allowed in an examination and you should aim for three within $0.10\,cm^3$.

Weighing

Use the same balance for all weighings and so eliminate any zero error that the balance may have. Ensure that the pan is clean before you start to use it. Read the mass to the appropriate number of significant figures, including a final zero (0) if necessary. If the reading is 8.560, the final zero should be included to show the accuracy.

Apparatus with gradations

Make sure that the gradation mark is horizontal and that your eye is also looking across it horizontally. Give yourself time to take the readings. If a reading lies between two gradations, decide whether it is nearer one or the other, or is nearer the mid-point, in which case record it. For example, if the reading on a $50\,cm^3$ measuring cylinder lies midway between 33 and $34\,cm^3$, record it as $33.5\,cm^3$.

Enthalpy change experiments

These are normally conducted in a calorimeter made from polystyrene (which has low heat conductivity). There is always some heat exchange with the surroundings when the temperature changes from room temperature. A temperature correction can be made graphically (see Chapter 6) by following the rise or fall in temperature over time and so making an adjustment to the maximum or minimum temperature reached.

Qualitative observations and recordings

All observations should be noted at the time the test is carried out. Before starting the experimental work, draw up a table, if it hasn't been done for you, to show both the test and the observations made.

Observations from inorganic or organic 'wet tests' should be recorded with a clear indication of the tests or practical procedures which led to the observations. A wet test is a method of analysis which involves making observations on a reaction carried out using a liquid phase. Titration and test-tube reactions using solutions are examples.

- In inorganic chemistry wet tests, it is important to record whether just a few drops or an excess of reagent were added as the outcome in the two situations can be different.
- Organic tests should be recorded with an indication of the test and the observation.

Ensure you record *observations* at this stage and not *conclusions* such as 'gas produced' or 'gets oxidised' – these are inferences from observations, not the observations themselves. In these cases, the observation might be of effervescence, a vapour, a distinctive smell or a colour change.

Note down:

- the colour of a solid or solution both before and after a reaction (note that a **clear** solution can be any colour; if there is no colour, then a **colourless solution** is the correct term to use)
- whether bubbles or effervescence are noticeable, whether a solid dissolves or whether a precipitate is formed
- if a solution gets warmer or colder.

On testing with a few drops of sodium hydroxide solution or ammonia solution, note:

- any colour change (state the colour both *before* and *after* the reaction)
- formation of any precipitate and its colour (take care when observing a precipitate formed in a coloured solution – allow the precipitate to settle before deciding on the colour).

On further testing with excess sodium hydroxide solution or ammonia solution, note:

- any further colour change
- any precipitate dissolving or remaining.

On leaving a mixture for a few minutes, note if there is any further change.

If bubbles are produced in a test, further tests need to be carried out to identify the gas; both the test and the observation need to be recorded.

You need to be familiar with carrying out the specified tests on anions and cations and tests on gases from the inorganic areas of the specification and be able to cite the expected observations from these (Tables 8.1, 8.2 and 8.3). You should also be familiar with carrying out tests using Tollens' reagent and Fehling's reagent, the tri-iodomethane test and oxidation reactions using potassium manganate(VII), and be able to cite the observations for positive identifications.

In your experimental report, you need to

- state where in a series of tests an observation has been made
- use the correct name or formula for any reagent selected for use in a test
- ensure that a test which shows no observational change is reported as such and not left blank.

Once the data has been collected and recorded, then it may be used to help you form your conclusions and enable you to answer the questions.

Decisions on measurements or observations

Titrations

To calculate the **mean titre**, the rough result is discarded and the value of two identical titres is used:

- If there are no agreeing values, the average of titres within 0.05 cm^3 is used.
- If this is still not possible, the average of titres which agree to within 0.1 cm^3 is used.
- Be clear which values you are using and how you have calculated the mean titre.
- The mean should be expressed to two decimal places, rounded appropriately.

Using equipment for measurement

You need to be able to identify where a repeat measurement or experiment is necessary. A value calculated from one measurement of temperature or mass is subject to the uncertainty of that one measurement and should always be repeated if time and amount of chemicals allow.

If a graph is plotted from a succession of values, as in a rate experiment, the graph itself may give an indication of anomalous values, which can be highlighted. Anomalous data or points on a graph should not be used when doing calculations or drawing lines of best fit.

Each piece of equipment gives its own uncertainty in the measured value; you cannot make your final value more precise than the percentage uncertainty in the equipment. Work out the percentage uncertainty from each piece of equipment used in an experiment and identify where the most uncertainty arises.

EXAMPLE 3.3

To find the enthalpy change, or heat, of solution when a solid salt is dissolved in water, the temperatures of the water at the start and of the solution at the end are measured to find the temperature change.

Each temperature is measured to 0.5 °C, which leads to a total uncertainty in the measurement of 1.0 °C. If the temperature change is measured to be 5.0 °C, then the uncertainty is $(1.0/5.0) \times 100\% = 20\%$. This leads to an uncertainty in the final value of 20%.

The mass may be known to three decimal places; a mass of 0.782 g measured on the balance is known to 0.0005 g and the uncertainty is $(0.0005/0.782) \times 100\% = 0.06\%$. This is much less than the uncertainty due to the temperature measurement.

You may conclude from the above example that a much larger temperature change would reduce the uncertainty in the final answer.

Qualitative observations

When observing what happens with heating a solid or mixing solutions, repeating a test may be helpful to clarify what occurred. The analysis tables (Tables 8.1, 8.2 and 8.3) indicate expected results and these enable you to select reagents to confirm ions present.

AS LEVEL

4 Practice questions

Quantitative experiments

1 a You are provided with 2.5 g of pure anhydrous sodium carbonate powder, Na_2CO_3, and all necessary volumetric glassware, as well as a top-pan balance and normal laboratory equipment. Describe, with all practical details, how you would make a standard solution of sodium carbonate using a 250 cm^3 volumetric flask.

Ensure you produce a clear, logical order of working which could be followed by another student. [6]

....

....

....

....

....

....

....

....

b Describe in detail the practical steps you would take to use the standard sodium carbonate solution to find the concentration of a hydrochloric acid solution which is approximately 1.0 mol dm^{-3}. The indicator bromophenol blue, which turns from yellow in acid to blue, is available.

Produce a clear, logical procedure which could be followed by another student to get to the same result. [6]

....

....

....

....

....

....

....

....

[Total: 12]

2 Acidified potassium manganate(VII) can be used to find the concentration of a hydrogen peroxide solution.

$$2MnO_4^-(aq) + 6H^+(aq) + 5H_2O_2(aq) \rightarrow 2Mn^{2+}(aq) + 8H_2O(l) + 5O_2(g)$$

10.0 cm^3 of a solution of hydrogen peroxide is pipetted into a conical flask and is then titrated with a 0.0500 mol dm^{-3} solution of $KMnO_4$ which is acidified with sulfuric acid.

The results are shown in the figure.

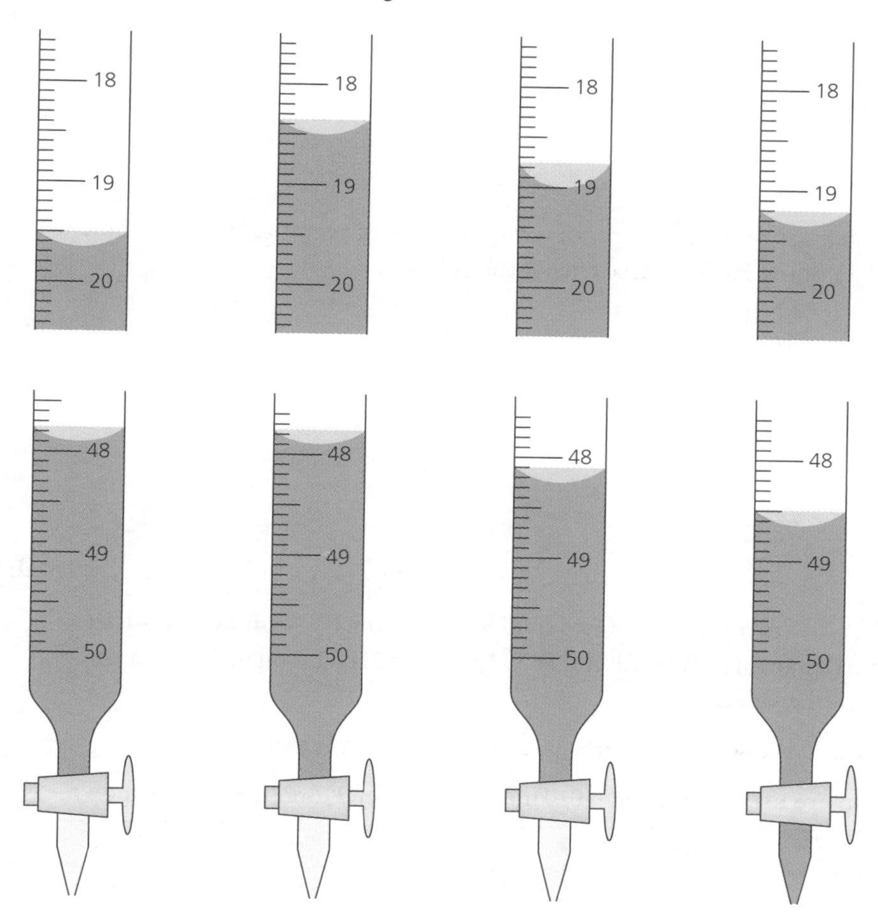

a Complete the table, noting that KMnO4 is in the burette. [2]

Titration	rough	1	2	3
Initial burette reading/cm^3				
Final burette reading/cm^3				
Volume $KMnO_4$ run in /cm^3				

In calculations, show all your working.

b Calculate the mean titre.

Mean titre .... [2]

c i Calculate the number of moles of $KMnO_4$ contained in the mean titre. [1]

ii Calculate the concentration of the H_2O_2 solution. [2]

d Calculate the volume of oxygen gas produced at room temperature and pressure by the complete decomposition to O_2 and H_2O of the H_2O_2 in 10.0 cm^3 of the solution. [2]

[Total: 9]

3 The greenish crystalline solid, **A**, is a mixture of the two double salts, iron(II) ammonium sulfate, $(NH_4)_2Fe(SO_4)_2.6H_2O$ and iron(III) ammonium sulfate, $FeNH_4(SO_4)_2.12H_2O$. You are to find the percentage by mass of the iron in the solid.

5.00 g of **A** is dissolved in water and the solution made up to 100 cm^3 in a volumetric flask.

10.0 cm^3 portions of the solution are titrated against 0.010 mol dm^{-3} $KMnO_4$ and the results shown.

	Rough	1	2	3
Final burette reading/cm^3	10.95	18.30	25.60	33.00
Initial burette reading/cm^3	3.50	10.95	18.30	25.60
Volume titre/cm^3				

a Show your working and work out the average titre to use in your calculations. [1]

....

....

b The portions are removed from near the bottom of the volumetric flask by pipette. What would be the effect on the titres if the volumetric flask had not been shaken sufficiently? [2]

....

....

c Work out the number of moles of MnO_4^- in your average titre. [1]

....

....

d Write the ionic equation for the oxidation of Fe^{2+} by MnO_4^-. [2]

....

....

e How many moles of Fe^{2+} are in the 100 cm^3 of solution? [1]

....

....

f What is the mass of iron(II) ammonium sulfate in **A**? [1]

....

....

g How many moles of iron(III) ions are in the solution before titration? [1]

....

....

h What is the total mass of iron in the 5.00 g of **A**? [1]

....

....

i What is the percentage of iron by mass in **A**? [1]

....

....

[Total: 11]

4 White crystalline potassium tetraoxalate has the formula $[(COOH)_2.(HOOCCOOK)].\mathbf{x}H_2O$. The number of molecules of water of crystallisation, **x**, can be found by titration with acidified $KMnO_4$. The reaction is slow at room temperature.

1.20 g of potassium tetraoxalate crystals are dissolved in water and the solution accurately made up to 250 cm^3 in a volumetric flask. 25.0 cm^3 portions of this solution are pipetted into a beaker and approximately 25 cm^3 of 2 mol dm^{-3} H_2SO_4 added. This solution is then warmed to between 60 °C and 70 °C and is then titrated with 0.012 mol dm^{-3} $KMnO_4$ in the burette. The first few drops added react slowly but then the reaction is fast – this is an example of autocatalysis in which a product catalyses the reaction.

The titration is repeated until three accurate titres within 0.10 cm^3 are obtained and the average titre calculated.

The average titre = 31.50 cm^3.

a Write the ionic equation for the oxidation of potassium tetraoxalate by acidified MnO_4^- to form CO_2, Mn^{2+}, K^+, and H_2O. [2]

....

....

b What would happen to the average titre if the solutions were not heated sufficiently? [1]

....

....

c Suggest what might happen to the titres if the solutions were heated to boiling. [1]

....

....

d State the colour change at the end point. [1]

....

....

e Calculate the number of moles of $KMnO_4$ in the average titre. [1]

....

....

f Calculate the number of moles of potassium salt in the 25.0 cm^3 pipetted. [1]

....

....

g How many moles of potassium salt are in the whole 250 cm^3 of solution? [1]

....

....

h From your answer to g and the mass of salt dissolved, calculate the relative formula mass of the salt. [1]

....

....

i Calculate the value of x. [1]

....

....

[Total: 10]

5 Nitrous acid, HNO_2, is a volatile weak acid which forms salts called nitrites. Nitrous acid and the nitrites can be oxidised by acidified potassium permanganate. The oxidation reaction is slow.

a Write a balanced equation for the oxidation of the nitrite ion, NO_2^-, by acidified MnO_4^- to form NO_3^-. [2]

....

....

b i 1.05 g $NaNO_2$ is dissolved in water and the solution then made up to 250 cm^3 in a volumetric flask.

25 cm^3 of this solution is pipetted into a second 250 cm^3 volumetric flask and the solution made up to the mark and thoroughly shaken.

Calculate the concentration of the diluted solution. [2]

....

....

ii 25 cm^3 of the diluted solution of $NaNO_2$ is pipetted into a conical flask and warmed.

A solution of acidified $KMnO_4$ with a concentration of 0.01 mol dm^{-3} MnO_4^- and 0.5 mol dm^{-3} H_2SO_4 is titrated into the warm $NaNO_2$ solution.

Calculate the volume of acidified $KMnO_4$ required to reach the end point. [3]

....

....

iii What is the purpose of warming the solution? [1]

....

....

iv What would be the effect of using 0.01 mol dm^- $KMnO_4$ in the burette and adding excess H_2SO_4 to the $NaNO_2$ before warming the mixture? [2]

....

....

v This method of adding acidified $KMnO_4$ to the $NaNO_2$ solution gives a value of the titre which is far too low. Suggest why this is so. [2]

....

....

c The following is a more accurate method for finding the concentration of $NaNO_2$ using titration with potassium permanganate. It involves a back titration; this means adding a known excess of a reagent and then finding how much is left over by titration.

Solid X is an impure sample of sodium nitrite, $NaNO_2$.

2.85 g of X is dissolved in water, the solution made up accurately to 250 cm^3 and 25 cm^3 of this solution is pipetted into a conical flask.

This solution is warmed and 100 cm^3 (an excess) of 0.0200 mol dm^{-3} $KMnO_4$ acidified with 0.5 mol dm^{-3} H_2SO_4 are added.

20 cm^3 of 0.100 mol dm^{-3} oxalic acid solution is now added and the mixture warmed. This reacts with the excess $KMnO_4$ but leaves an excess of oxalic acid.

The mixture is titrated with 0.0200 mol dm^{-3} acidified $KMnO_4$ and the titre noted.

The procedure is repeated to get three titres agreeing to within 0.10 cm^3 and the average titre volume calculated.

Average titre = 8.00 cm^3.

i Calculate the number of moles of $KMnO_4$ in the final titration. [1]

....

....

ii Write the equation for the reaction of oxalic acid with MnO_4^-. [2]

....

....

iii Calculate the number of moles of 0.100 mol dm^{-3} oxalic acid which react with the 8.00 cm^3 $KMnO_4$ in the final titre. This is the excess oxalic acid added. [1]

....

....

iv Calculate the total number of moles of oxalic acid added and hence the number of moles of oxalic acid which had reacted with the excess $KMnO_4$ initially added to the solution of X. [2]

....

....

v Calculate the excess number of moles of $KMnO_4$ initially added. [1]

....

....

vi Calculate the number of moles of 0.0200 mol dm^{-3} $KMnO_4$ initially added and use this value to calculate the number of moles of $KMnO_4$ which reacted with the 25 cm^3 of $NaNO_2$ solution. [2]

....

....

vii Use your equation in a to calculate the number of moles of $NaNO_2$ in 25 cm³ of solution. [1]

....

....

viii Calculate the number of moles and hence the mass of $NaNO_2$ in 2.85 g of X. [1]

....

....

ix Calculate the percentage purity of X. [2]

....

....

[Total: 25]

6 A 3.15 g sample of impure copper sulfate crystals, $CuSO_4.5H_2O$, is dissolved in water and the solution made up accurately to 250 cm³.

a Describe how to make this solution. [4]

....

....

....

....

....

....

A 25 cm³ sample of the copper sulfate solution is pipetted into a conical flask and excess potassium iodide solution added.

$$2CuSO_4(aq) + 4KI(aq) \rightarrow 2CuI(s) + 2K_2SO_4(aq) + I_2(aq)$$

b Describe what you would observe on the addition of the potassium iodide solution. [3]

....

....

....

The solution is titrated with 0.051 mol dm^{-3} sodium thiosulfate, $Na_2S_2O_3$.

$$I_2(aq) + 2S_2O_3^{2-}(aq) \rightarrow 2I^-(aq) + S_4O_6^{2-}(aq)$$

c Describe how the titration is carried out. [6]

....

....

....

....

....

....

The burette readings for four titration are shown in the table.

	1	2	3	4
	0.50	1.20	3.45	24.00
	23.25	21.60	23.95	44.50

d **i** Complete the table, including the headings. [4]

ii Select suitable results and calculate the mean titre. [1]

iii Using the mean titre, calculate the concentration of the copper sulfate solution. [3]

iv Calculate the mass of $CuSO_4.5H_2O$ in the crystals. [3]

v Calculate the percentage purity of the copper sulfate crystals. [2]

[Total: 26]

7 In neutral solution, iodide ions catalyse the decomposition of hydrogen peroxide.

$$2H_2O_2(aq) \rightarrow 2H_2O(l) + O_2(g)$$

The rate of the reaction can be monitored by measuring the volume of oxygen produced at known time intervals.

In this experiment, 10.0 cm^3 of 0.100 mol dm^{-3} KI and 15.0 cm^3 of deionised water are added to a conical flask which acts as the reaction vessel. The oxygen is collected over water in an inverted burette. The initial burette reading is noted and the temperature is measured. A stopwatch is started as 5.0 cm^3 of 0.880 mol dm^{-3} H_2O_2 is added to the conical flask.

a Draw a labelled diagram of the apparatus. [4]

A student made the following measurements at a temperature of 20 °C.

Burette reading/cm^3	48.70	47.70	46.70	45.70	44.70	43.70	42.70	41.70
Time/s	0	35	75	115	155	193	231	270

Burette reading/cm^3	40.70	39.70	38.70	37.70	36.70	35.70	34.70	33.70
Time/s	310	358	403	445	498	550	612	690

b Draw a table to show the total volume of oxygen released at each time measurement. [2]

c **i** Plot a graph showing the volume of oxygen released against the time. [4]

ii What volume of oxygen has been released after 300 s? [1]

....

iii Work out the average rate of reaction in $cm^3 s^{-1}$ for the first 300 s. [2]

iv Find the rate of reaction at $t = 600$ s by drawing a tangent to the curve at this point. Evaluate the gradient of the tangent. [3]

v State and explain the difference between the two values in **iii** and **iv**. [2]

....

....

....

....

d On the same axes, sketch the graph you would expect to get if the reaction had been carried out at 25 °C. [2]

e Explain how you would show that the iodide ions act as a catalyst in this decomposition. [2]

....

....

....

....

....

....

....

....

[Total: 22]

8 A measuring cylinder is used to measure 50 cm^3 of 1 mol dm^{-3} NaOH which is placed into a 200 cm^3 beaker. A thermometer is placed into the beaker and the temperature, T_0, noted. 50 cm^3 of 1 mol dm^{-3} CH_3CO_2H is added and the mixture stirred. The temperature of the mixture is taken every 20 seconds from the time of the addition. The following results are obtained.

Time/s	0	20	40	60	80	100	120	140	160	180	200	240	280
Temperature/°C	19.0	20.5	21.5	23.0	24.0	24.0	23.5	23.5	23.0	23.0	22.5	22.0	21.5

a i Plot a graph showing how the temperature varies with time. [4]

ii Extrapolate the cooling curve to find the maximum temperature, T_{max}, which would be achieved by the mixture if no heat was lost from the apparatus. [1]

iii Using your value of T_{max}, calculate the temperature change of the solution, ΔT_{max}, if no heat had been lost from the apparatus. [2]

iv Calculate $\Delta H_{neutralisation}$ for this reaction. (specific heat capacity of water = $4.18\,J\,g^{-1}\,K^{-1}$) [4]

b The accepted value for the amount of heat released during a neutralisation is $57.9\,kJ\,mol^{-1}$. Assuming the heat loss from the apparatus has been taken into account, suggest a reason for any difference between this value and your calculated value. [2]

....

....

....

c What improvements could you make to the procedure to obtain a more accurate value for the enthalpy of neutralisation? [1]

....

....

....

[Total: 14]

9 A sample of zinc carbonate is contaminated with zinc sulfate. To find the percentage purity by mass of the sample, two methods are used.

Method 1

A crucible is weighed and then a portion of the contaminated sample is added and the crucible and contents weighed again. The crucible and contents are then heated strongly over a Bunsen flame for 5 minutes and then left to cool. The crucible and contents are weighed again.

- Initial mass of crucible = 74.58 g
- Initial mass of crucible and contents = 76.08 g
- Final mass of crucible and contents = 75.70 g

a **i** Calculate the mass of contaminated sample taken. [1]

ii Calculate the loss in mass of the sample on heating and hence the number of moles of $CO_2(g)$ lost. [2]

iii State the number of moles of $ZnCO_3$ in the sample. [1]

iv Calculate the percentage by mass of $ZnCO_3$ in the sample. [2]

Method 2

1.50 g of the contaminated mixture is dissolved in 100 cm^3 of 0.5 mol dm^{-3} HCl in a beaker and then the solution made up to 250 cm^3 using a volumetric flask.

$$ZnCO_3(s) + 2HCl(aq) \rightarrow ZnCl_2(aq) + H_2O(l) + CO_2(g)$$

25 cm^3 aliquots of the solution are pipetted and titrated against 0.100 mol dm^{-3} NaOH(aq) using thymol blue as indicator.

	1	2	3	4
Initial burette reading/cm^3	1.30	2.40	3.55	2.75
Final burette reading/cm^3	32.25	32.50	33.70	32.95
Volume NaOH(aq) run in/cm^3				

b Complete the table. [2]

c **i** What is the colour change seen at the end-point? [1]

....

ii Work out the mean titre, indicating clearly which values you are using. [2]

iii Work out the number of moles of NaOH required to neutralise the HCl in a 25 cm^3 aliquot of solution. [1]

iv Work out the number of moles of HCl which have reacted with the $ZnCO_3$ in 1.50 g of the mixture. [3]

v Work out the mass of $ZnCO_3$ in 1.50 g of mixture. [2]

vi Calculate the percentage by mass of $ZnCO_3$ in the mixture. [2]

d i Identify errors introduced in the two methods and explain how these can be minimised. [2]

....

....

....

....

ii Which of the two methods leads to the more accurate value for the percentage? [1]

....

[Total: 22]

10 Acidified potassium iodide reacts with hydrogen peroxide to produce iodine and the rate of reaction can be monitored by the colour change.

$$2I^-(aq) + 2H^+(aq) + H_2O_2(aq) \rightarrow I_2(aq) + 2H_2O(l)$$

A small amount of sodium thiosulfate is added and reacts with the iodine produced.

$$2S_2O_3^{2-}(aq) + I_2(aq) \rightarrow S_4O_6^{2-}(aq) + 2I^-(aq)$$

When all the thiosulfate has been used up, the iodine is no longer removed and colours the solution.

To make the colour change more obvious, a small amount of starch is added to the mixture. The starch forms an intense blue colour with the iodine. The time taken from adding the reagents to the appearance of the blue colour is a measure of the rate of reaction.

Identify any systematic errors in the method and explain how you would minimise them. [4]

....

....

....

....

....

....

....

[Total: 4]

11 *If your teachers are able to prepare the practical experiment according to the instructions given for the appropriate examination questions in June 2016, carry out the experiment and use your own results.*

If this is not possible, use the following results to calculate and record the mass of FA1 added to the acid and the mass of CO2 given off and use these values in the calculations in Q12(b).

Mass of weighing bottle + FA1/g	11.78
Mass of weighing bottle + residue/g	10.01
Mass of beaker + 25 cm³ HCl/g	165.53
Mass of beaker + acid + FA1 after reaction/g	166.31

In this experiment you will determine the identity of the Group 2 metal, **X**, in the carbonate, $\mathbf{XCO_3}$. To do this you will react a known mass of $\mathbf{XCO_3}$ with **excess** hydrochloric acid, HCl, and measure the mass of carbon dioxide that is given off.

- **FA 1** is XCO_3.
- **FA 2** is hydrochloric acid, HCl.

a Method

- Weigh the stoppered tube containing **FA 1** and record its mass.
- Use the measuring cylinder to transfer 25 cm³ of **FA 2** into the 250 cm³ beaker.
- Weigh the beaker containing the acid and record the mass.
- Carefully add all the sample of **FA 1** to the acid in the beaker.
- Stir the mixture until there is no further reaction.
- Reweigh the beaker and its contents and record the mass.
- **Keep the contents of the beaker for use in question 10.**
- Reweigh the stoppered tube containing any residual **FA 1** and record its mass.
- Calculate the mass of **FA 1** added to the acid and record this value.
- Calculate the mass of carbon dioxide given off and record this value.

[7]

b Calculations

Show your working and appropriate significant figures in the final answer to each step of your calculations.

i Calculate the number of moles of carbon dioxide given off when **X**CO_3 reacted with the acid.

Use the data in the periodic table on page 95.

Moles of CO_2 = ... mol

ii Write the equation for the reaction of **FA 1**, **X**CO_3, with hydrochloric acid, HCl. Include state symbols.

....

iii Use your answers to **i** and **ii** to calculate the number of moles of **X**CO_3 that were added to the acid.

Moles of **X**CO_3 = ... mol

iv Use your answer to **iii** to calculate the relative atomic mass, A_r, of **X**.

Identify **X**.

A_r of **X** = ...

X is.......................... [5]

c One of the sources of error in this experiment is that it is very difficult to limit acid spraying out of the beaker when the metal carbonate is added to the acid.

i Explain what effect this acid spray would have on the value you calculated for the relative atomic mass, A_r, of **X**.

....

....

....

ii Why is a small amount of acid spray not likely to cause an error in the identification of **X**?

....

....

....

iii How could you minimise acid spraying out of the beaker?

....

....

....

[3]

[Total: 15]

Cambridge International AS & A Level Chemistry 9701 Paper 31 Q1 June 2016

12 In this experiment you will determine the concentration of the hydrochloric acid, **FA 2**, used in **Question 9**. You will first dilute the reaction mixture that you prepared in **Question 9** and then titrate this diluted solution against sodium hydroxide, NaOH.

If you are not able to carry out the titration, use the following values to answer the question.

Final burette reading/cm^3	39.15	38.50	37.85	41.10
Initial burette reading/cm^3	2.45	1.75	1.30	4.55
Titre/cm^3				

$$HCl\,(aq) + NaOH(aq) \rightarrow NaCl\,(aq) + H_2O(l)$$

FA 3 is 0.0400 mol dm^{-3} sodium hydroxide, NaOH.
Methyl orange indicator

a **Method**

Dilution

- Transfer all the reaction mixture that you prepared in **9a** from the 250 cm^3 beaker to the 250 cm^3 volumetric flask.
- Rinse the beaker with a little distilled water and add these washings to the volumetric flask.
- Fill the volumetric flask to the line with distilled water. Stopper the flask and shake it to ensure thorough mixing.
- Label this solution **FA 4**.

Titration

- Fill the burette with **FA 4**.
- Use a pipette to transfer 25.0 cm^3 of **FA 3** into a conical flask.
- Add a few drops of methyl orange.
- Perform a rough titration and record your burette readings in the space below.

The rough titre is .... cm^3.

- Carry out as many accurate titrations as you think necessary to obtain consistent results.
- Make certain any recorded results show the precision of your practical work.
- Record in a suitable form below all of your burette readings and the volume of **FA 4** added in each accurate titration.

[4]

b From your accurate titration results, obtain a suitable value for the volume of **FA 4** to be used in your calculations. Show clearly how you obtained this value.

$25.0\,cm^3$ of **FA 3** required .... cm^3 of **FA 4**. [1]

c Calculations

Show your working and appropriate significant figures in the final answer to each step of your calculations.

i Calculate the number of moles of sodium hydroxide, NaOH, present in $25.0\,cm^3$ of **FA 3**.

Moles of NaOH = ... mol

ii Calculate the number of moles of hydrochloric acid, HCl, present in $250\,cm^3$ of **FA 4**.

Moles of HCl in $250\,cm^3$ of **FA 4** = ... mol

iii Use your answers to **9b i** and **9b ii** to calculate the number of moles of HCl that reacted with **FA 1** in the experiment you carried out in **Question 9**.

Moles of HCl that reacted with **FA 1** = ... mol

iv Use your answers to **10c ii** and **10c iii** to calculate the concentration of **FA 2**.

Concentration of **FA 2** = ... $mol\,dm^{-3}$ [5]

d i One of the sources of error in determining the concentration of **FA 2** involves measuring volumes of solutions in both **Questions 9** and **10**.

State which volume of solution that you have measured has the greatest percentage error.

How could you have reduced this error?

....

....

....

ii A student suggested that a greater mass of **XCO_3** should be used so that the average titre calculated in **10b** would be a greater volume.

Explain whether you agree with the student that this would lead to a greater volume for the average titre. [2]

....

....

....

....

[Total: 12]

Cambridge International AS & A Level Chemistry 9701 Paper 31 Q2 June 2016

Observational experiments

13 Describe what would be seen when the following reactions are carried out.

a Sodium hydroxide solution is added a few drops at a time until in excess to copper sulfate solution. [2]

....

....

b Ammonia solution is added a few drops at a time until in excess to copper sulfate solution. [2]

....

....

c Zinc carbonate crystals are added to aqueous hydrochloric acid. [2]

....

....

d Silver nitrate solution acidified with nitric acid is added to a mixture of potassium chloride and potassium iodide in aqueous solution and then concentrated ammonia solution is added to the mixture. [3]

....

....

e Powdered manganese dioxide is stirred into an aqueous solution of hydrogen peroxide. [2]

....

....

f Aqueous ammonia is added to a mixture of iron(III) sulfate and copper(II) sulfate in aqueous solution. [3]

....

....

g Ten drops of cyclohexene are added to a few drops of an aqueous solution of $KMnO_4$ acidified with $1\,mol\,dm^{-3}$ sulfuric acid. The mixture is shaken. [2]

....

....

h Aqueous ammonia is added to about $2\,cm^3$ aqueous silver nitrate until the precipitate formed just dissolves. Ten drops of benzaldehyde, C_6H_5CHO, are added to the mixture which is then warmed in a water bath. [1]

....

....

i Ten drops of propanal are added to $2\,cm^3$ of Fehling's solution and heated in a water bath. [1]

....

....

j Aqueous sodium hydroxide is added drop by drop to $1\,cm^3$ of aqueous iodine until the mixture is nearly colourless. Five drops of propanone are added and the mixture warmed in a water bath for just 1 minute before being removed and left to cool. [1]

....

....

[Total: 19]

14 a A solution containing both $FeCl_2$ and $FeSO_4$ is tested separately with NaOH(aq), $BaCl_2$(aq) and acidified $KMnO_4$.

Draw up a table showing the three tests and the observations you would expect to make. [3]

b A solution contains both $CuSO_4$ and $Al_2(SO_4)_3$.

Draw up a table showing the observations you would expect to make if portions of this solution are tested separately with NaOH(aq) and NH_3(aq). [5]

[Total: 8]

15 The powder **X** contains two cations and one anion. It is heated and the gas evolved is passed through lime water causing a white precipitate.

A sample of **X** is dissolved in HCl(aq) and the solution used for the following tests:

- NaOH(aq) is added dropwise to a portion of the solution and a white precipitate is seen which remains on the addition of excess NaOH(aq).
- Warming the alkaline mixture obtained gives fumes which turn moist red litmus blue.

a Show these results in a suitable table. [2]

b What cations are present in **X**? Explain your choice. [2]

....

....

....

....

c What anion is present in **X**? Explain your choice. [1]

....

....

d What observations would you make if a small portion of **X** is stirred into pure water? What differences in the observations would there be if $H_2SO_4(aq)$ was used instead of pure water? Explain your answer. [4]

....

....

....

....

....

....

[Total: 9]

16 A student made the following observations when some reactions were carried out on separate portions of solution **Y**:

- adding a few drops of NaOH(aq) produces a green precipitate
- adding excess NaOH(aq) produces a green precipitate in a green solution; on setting the mixture to one side, the green precipitate starts to turn brown
- adding a few drops of $NH_3(aq)$ produces a green precipitate
- adding excess $NH_3(aq)$ does not produce any further changes but leaving the mixture to one side, the precipitate starts to turn brown
- two drops of acidified $KMnO_4$ added to solution **Y** are decolourised.

a What type of reaction is the reaction with acidified $KMnO_4$? [1]

....

b Identify the cations in solution **Y** and explain the reasons for your choice. [6]

....

....

....

....

....

....

....

....

c Explain why the green precipitate turns brown. [2]

....

....

d What could be used to show that the anion is sulfate and what observations would be made? [3]

....

....

....

e Write a correctly balanced equation for the reaction between $MnO_4^-(aq)$ and an ion in solution **Y**. [2]

....

[Total: 14]

17 This practical experiment involved observations using **X** and **Y**.

- **X** is a black powder.
- **Y** is a colourless solution.

	Test	Observation
1	A small spatula of **X** is stirred into half a test-tube of water and then left for several minutes.	Black solid sinks to the bottom of the test-tube.
2	A small spatula of **X** is stirred into half a test-tube of **Y**.	Effervescence occurs. No colour change is observed. The black solid remains at the bottom of the test-tube.
3.1	Acidified $KMnO_4$ is added dropwise to half a test-tube of **Y**.	Effervescence occurs. The $KMnO_4$ is decolourised.
3.2	Further acidified $KMnO_4$ is added dropwise to the test-tube of mixture until no more bubbles are produced. NaOH(aq) is then added until the solution is alkaline.	Off-white precipitate forms.
3.3	Further NaOH(aq) is then added until in excess.	No further change occurs.
3.4	**Y** is added to the off-white precipitate from test 3.3.	The off-white precipitate turns black.
4	To a new portion of **Y** in a test-tube, neutral $KMnO_4$(aq) is added dropwise and slowly.	The $KMnO_4$ is decolourised and a black precipitate forms. Effervescence occurs.

a i Explain at what point in the practical procedure you would test for carbon dioxide and how this would be carried out. [2]

....

....

ii Any gas produced was tested with moist red litmus paper but there was no colour change. What information does this give about the nature of the gas? [1]

....

....

iii All gases were tested with a lighted splint which continued to burn more brightly in every case. Suggest what gas is produced. [2]

....

....

b i From the evidence in test 3.1, what kind of reagent is **Y**? [1]

....

....

ii Name the off-white precipitate produced in test 3.2 and write an equation for its formation. [2]

....

....

iii Explain what has happened to the $KMnO_4$ in test 4 and suggest the formula of the black precipitate. [2]

....

....

iv What kind of reaction turns the off-white precipitate to black in test 3.4? Explain your answer by suggesting the reagent and product responsible for the colour change. [2]

....

....

v Suggest the formula of the compound in **Y** and write an equation for the reaction occurring in test 3.4. [3]

....

....

[Total: 15]

18 *If your teacher and laboratory assistants are able to prepare this practical experiment, according to the Special Instructions sent to schools in June 2018, carry out the experiment as a helpful learning experience for working under examination conditions.*

If the experiment cannot be prepared:

- *assume FA 5 is a mixture of propanol and propanoic acid and answer 20a with what you would expect to see.*
- *assume FA 7 is $CuCO_3$ and FA 8 is $Zn(NO_3)_2$ and answer 20b i with what you would expect to see from these tests and in 20b iii explain what other possibilities there might be from the observations in 20b i and 20b ii.*

Where reagents are selected for use in a test, the **name** or **correct formula** of the element or compound must be given.

At each stage of any test you are to record details of the following:

- colour changes seen
- the formation of any precipitate and its solubility in an excess of the reagent added
- the formation of any gas and its identification by a suitable test.

You should indicate clearly at what stage in a test a change occurs.

If any solution is warmed, a **boiling tube** must be used.

Rinse and reuse test-tubes and boiling tubes where possible.

No additional tests for ions present should be attempted.

a Half fill the 250 cm^3 beaker with water. Heat to approximately 70 °C, then turn off the Bunsen burner. This will be used as a water bath.

i **FA 5** is an aqueous solution of an organic compound. Carry out the following tests on **FA 5** and record your observations in the table. [4]

Test	Observations
To a 1 cm depth of **FA 5** in a test-tube add a small spatula measure of sodium carbonate.	
To a 1 cm depth of **FA 5** in a test-tube add two drops of acidified potassium manganate(VII). Leave to stand in the water bath.	
To a 1 cm depth of **FA 5** in a test-tube add a few drops of aqueous silver nitrate.	
To a 1 cm depth of aqueous silver nitrate in a test-tube add a few drops of aqueous sodium hydroxide and then add aqueous ammonia slowly until the grey precipitate that forms **just** dissolves. This is Tollens' reagent. To this solution add a 1 cm depth of **FA 5** and leave to stand in the water bath. **Care: rinse the tube as soon as you have completed this test.**	

ii Suggest **two** functional groups that could be present in **FA 5**. [2]

....

....

b **FA 6** is a mixture that contains two cations and two anions from the Qualitative Analysis Notes. Distilled water was added to **FA 6**, the mixture was stirred and then filtered. You are provided with the dried residue, **FA 7**, and the filtrate, **FA 8**, from this process.

i **Tests on the residue, FA 7**

Carry out the following tests and record your observations in the table. [3]

Test	Observations
Place a spatula measure of **FA 7** in a boiling tube. Add dilute hydrochloric acid until no further reaction occurs, then	
transfer a 1 cm depth of the solution into a test-tube. To this add aqueous sodium hydroxide.	

ii **Tests on the filtrate, FA 8**

Carry out the following tests and record your observations in the table. [3]

Test	Observations
To a 1 cm depth of **FA 8** in a boiling tube add a 1 cm depth of aqueous sodium hydroxide, then	
warm gently.	
To a 1 cm depth of **FA 8** in a boiling tube add a piece of aluminium foil and a 1 cm depth of aqueous sodium hydroxide. Warm gently.	

iii Conclusions about cations

State **one** cation that is **definitely** present in **FA 6**.

State **two** possible identities for the other cation present in **FA 6**.

Suggest how you could determine which of these two possible cations is present.

Do not carry out this test. [3]

iv Conclusions about anions

State **one** anion that is **definitely** present in **FA 6**.

State **two** possible identities for the other anion present in **FA 6**. [2]

[Total: 17]

Cambridge International AS & A Level Chemistry 9701 Paper 31 Q1 June 2018

19 Some data from the containers of two water-soluble organic compounds, **A** and **B**, is given.

A; melting point = 336K;

B; boiling point 356K;

a What is the meaning of each of the labels? [2]

....

....

....

....

A and **B** could be the following:

- a carboxylic acid
- a secondary alcohol
- an aldehyde.

Results of tests on separate solutions of **A** and **B** are shown:

Test	A	B
Warm with acidified $K_2Cr_2O_7$	no change	orange to green
Addition of sodium carbonate solution	effervescence	no change
Warm with alkaline iodine solution	no change	yellow crystals
Addition of 2,4-dinitrophenylhydrazine	no change	no change

2,4-dinitrophenylhydrazine is a dangerous chemical and should not be used in practical activities.

b To which functional group could **A** and **B** be allocated? [5]

A .... **B** ....

Explain your choices.

....

....

....

....

The mass spectrum of **A** is shown; it indicates a second functional group in **A**.

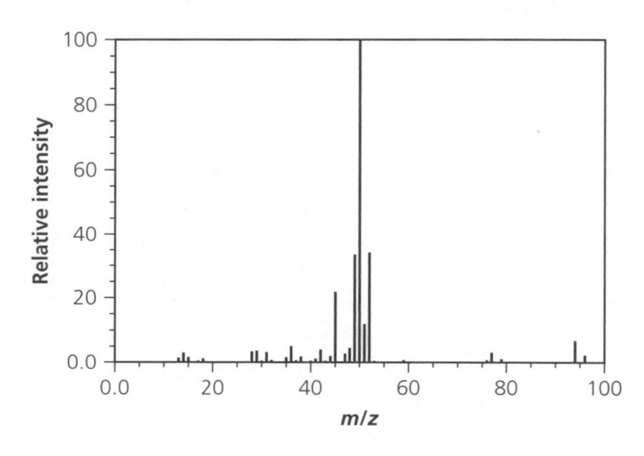

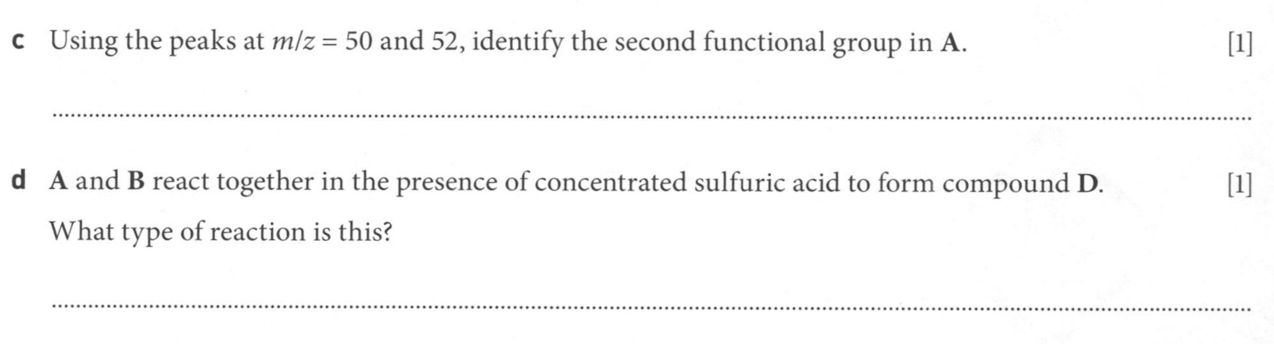

c Using the peaks at $m/z = 50$ and 52, identify the second functional group in **A**. [1]

....

d **A** and **B** react together in the presence of concentrated sulfuric acid to form compound **D**. [1]

What type of reaction is this?

....

[Total: 9]

A LEVEL

5 Skills

Defining the problem and giving the method

Defining the problem involves choosing an experimental method to safely and accurately answer a chemical question such as:

- What is the concentration of … ?
- What is the formula of … ?
- What is the enthalpy change of … ?
- What is the rate of … ?
- How would you prepare a sample of … ?
- How do you test the purity of … ?

You may need to provide full details of a chemical method to answer one of these questions. Your procedure would need to be safe and efficient, with no unnecessary steps and should lead to a reliable result. This means the result should answer the chemical question being asked, rather than just gathering data.

You should be able to list the steps needed to carry out your procedure, identifying suitable chemical apparatus required, and be able to identify any risks that would need to be controlled, for example avoiding the fire risk of heating an organic liquid with a Bunsen burner by using an electrical heater or water bath instead.

Your method could include a diagram of the apparatus to show how it is arranged and you may be asked to calculate the initial quantities of chemicals needed. For example, if you use a $100\,cm^3$ gas syringe to collect the oxygen released from the catalysed decomposition of hydrogen peroxide, you may have to calculate the volume of hydrogen peroxide, of a given concentration, to use. In this example, you would aim to release enough oxygen gas to get a reading on the gas syringe, without exceeding its $100\,cm^3$ capacity.

You should be able to identify the independent, dependent and control variables in an experiment.

- The **independent variable** is the variable that you decide on or change during the experiment.
- The **dependent variable** is the variable that changes as a result of you altering the independent variable.
- **Control variables** are variables you keep the same throughout the experiment, ensuring that the change in the dependent variable arises as a result of the change in independent variable only.

In an experiment to see how the concentration of a reactant affects the rate of reaction:

- the independent variable is the concentration of the reactant
- the dependent variable is the rate of reaction (or a measurement that results in determination of the rate of reaction)
- control variables may be the concentration of other reactants or the temperature of the experiment, for example.

You should be prepared to include in your method how you will vary the independent variable. In the rates example, this could be five different concentrations of the reactant, with details about how the concentrations of the other reactants remain the same (this is normally achieved using less of the reagent and water to keep the overall volume the same. This keeps the concentrations of the other reagents, which are control variables, constant.) You may also be asked to draw up a table in which to write the results or observations of your experiment. The table should include, as a minimum, columns for the independent variable and the dependent variable. Control variables are normally stated (as they should not change between experiments) but sometimes are, for instance the temperature of a reaction mixture if repeats are done on different days which would enable you to see if a change in temperature affects the dependent variable.

You should be able to express the aim of the experiment in the form of a prediction. In this example, it may be, 'The aim of the experiment is to show that the order of reaction with respect to (reagent) is first order.'

You should also be able to show the expected results in the form of a graph, normally as a sketch of the predicted shape of the graph rather than plotting data.

Finally, you should be able to show an understanding of why a method will be effective in answering the chemical question. In the rates experiment example above, this would mean being able to explain that any increase in the rate of reaction is due to the increase in concentration of the reactant being changed because all other reagents' concentrations remain constant.

The chemistry involved may not be part of the A Level syllabus. Where this occurs, enough information will be given to allow you to understand the practical aspects of the question you need to consider.

Analysis, conclusions and evaluation

Analysis of data

The analysis of data normally involves a series of calculations and/or graphs to determine a value or a relationship between two variables. The form these take depends on the problem being addressed.

To analyse data, you may have to:

- complete a table
- draw a graph
- calculate a series of concentrations.

For example, the activation energy for a reaction, E_A, can be found by measuring the rate constant for a reaction, k, at different temperatures. The relationship between the two is given by the Arrhenius equation:

$$k = Ae^{\left(\frac{-E_A}{RT}\right)}$$

which can, after taking the natural logs of each side, be rearranged to resemble the equation for a straight line:

$$y = \quad mx \quad + c$$

$$\ln(k) = -\frac{E_A}{R}\cdot\frac{1}{T} + \ln(A)$$

In this way, a graph of $\ln(k)$ against $1/T$ will have a gradient of $-E_A/R$.

You may be given rate constant data for several experiments at different temperatures and be asked to:

- calculate the temperatures in Kelvin rather than °C, then calculate $1/T$ for each
- calculate $\ln(k)$ from given values of k
- plot a graph of $\ln(k)$ against $1/T$ and measure the gradient
- calculate E_A from the gradient.

You may also need to suggest improvements to experimental methods. For example, if a titration was being conducted using a pH meter and measurements are taken every 2.0 cm^3 added from the burette, the end-point may be missed. The improvement would be to repeat the experiment, taking readings with every 0.2 cm^3 near the end-point. Taking readings with this frequency throughout the experiment would be impractical (if the end-point occurred at, say, 17.5 cm^3 added then more than 85 readings would have to be taken, most of them very similar. If the initial experiment showed a large pH change between 16.0 and 18.0 cm^3 added, the more frequent readings should be taken between these two volumes only when the experiment is repeated).

Finally, you may be asked to account for anomalies or errors in the data and suggest actions to mitigate them. For instance, if data includes an anomalous point you may be asked to suggest a reason for it and the desire to take additional readings to confirm or refute it. You may also be presented with systematic errors and invited to suggest improvements in the method to eliminate them. For instance, a consistently low temperature change on measuring an enthalpy change normally means insufficient insulation of the reaction vessel which could be improved by using a better insulated vessel or adding a lid.

Conclusion

Conclusions are normally short, justified statements about the chemistry, which are supported by the experimental results. They follow from the analysis of results, for example:

'As the rate of reaction is directly proportional to the concentration of reactant *x*, the order of reaction is 1st order with respect to reactant *x*.'

If this were related to the mechanism of the reaction, theory could be used to explain further.

'This is supported by the proposed mechanism, which involves one molecule of the reactant *x* reacting with two molecules of the other reactant *y* in the rate determining step. This means the reaction should be 1st order with respect to reactant *x* and second order with respect to reactant *y*.'

Further experiments could be suggested to extend the investigation.

'Further experiments in which the concentration of reactant *x* is a control variable and the concentration of reactant *y* is the independent variable should show that the rate of reaction is 2nd order with respect to reactant *y*.'

Evaluation

The critical analysis of an experimental method is often the most difficult part of the question. As well as discussing the reliability of experimental data, you may be presented with 'what if' questions. For these, you are expected to predict what effect a change or fault in the experiment will have on the final outcome.

For example, the concentration of an alkali is determined by titration with an acid. The titration involves placing a set volume of alkali into a conical flask with a pipette and titrating against an acid of known concentration. If slightly too much alkali was drawn into the pipette such that the level was above the graduation mark, what effect would this have on the concentration of alkali calculated?

To answer, you must consider that the volume of alkali transferred is too great, so the moles of acid used to reach the end-point would be too large and so conclude that the alkali will appear more concentrated than it actually is; that is, the calculated concentration would be higher than the actual concentration.

Answering questions of this type requires clear thought and the ability to predict how changes in the data affect the outcome of calculations.

A LEVEL

6 Methods

Finding a formula using mass changes

Mass changes on heating

Reactions which involve the loss of a gas (decomposition) or driving off water (waters of crystallisation) are often used to find a formula. For example, if a Group II carbonate decomposes, measuring the mass of starting material and the mass of the residue allows you to determine the relative formula mass of the carbonate and hence the formula.

The method is called **heating to a constant mass.**

1. Record the mass of a crucible.
2. Add the chemical to be tested and record the mass of the crucible and chemical.
3. Heat (gently at first, then strongly) the crucible and chemical.
4. Allow to cool, then record the mass of crucible and residue.
5. Repeat steps 3 and 4 until the masses recorded do not change from one measurement to the next.

The repeats ensure that the reaction is complete. From the measurements you can calculate the mass of chemical you start with and the mass of residue at the end. From these, you can determine the mass of gas given off. If you know the reaction products, this gives you the amount of gas in moles from which you can determine the number of moles of the chemical and hence its formula.

From your chemical knowledge, you should know whether a solid just loses waters of crystallisation on heating or decomposes as well. The experiment may still be useful, if you take the decomposition into account.

EXAMPLE 6.1

Magnesium nitrate will decompose at approximately 330 °C, a temperature easily achievable with a Bunsen burner. The full equation for the decomposition of the hydrated solid is:

$$2Mg(NO_3)_2.xH_2O(s) \rightarrow 2MgO(s) + 4NO_2(g) + O_2(g) + 2xH_2O(g)$$

A student made the measurements shown in the table.

Mass of crucible + lid/g	19.37
Mass of crucible + lid + hydrated magnesium nitrate/g	24.37
Mass of crucible + residue after 1st heating/g	20.72
Mass of crucible + residue after 2nd heating/g	20.20
Mass of crucible + residue after 3rd heating/g	20.17

Calculate the value of x.

Method

Calculate the mass of the $Mg(NO_3)_2.xH_2O$. This is the difference between the first two mass measurements.	$24.37 - 19.37 = 5.00\,g$
Calculate the mass of MgO at the end (this is the residue as the other two products are gases). The final two readings are close enough to each other to be considered a constant mass, so use the lower of the two.	$20.17 - 19.37 = 0.80\,g$ of MgO
Calculate the number of moles of MgO. Make sure you use the mass of Mg from your periodic table.	$0.80/(24.3 + 16) = 0.0199$ moles of MgO
This is also the number of moles of $Mg(NO_3)_2.xH_2O$. Calculate the mass of the $Mg(NO_3)_2$ contained in 0.0199 moles of $Mg(NO_3)_2.xH_2O$.	$148.3 \times 0.0199 = 2.95\,g$ of $Mg(NO_3)_2$
The rest of the mass of the $Mg(NO_3)_2.xH_2O$ is H_2O.	$5.00 - 2.95 = 2.05\,g$ of H_2O
Calculate the number of moles of H_2O.	$2.05/18 = 0.114$ moles of H_2O
So in the solid, 0.0199 moles of $Mg(NO_3)_2$ is combined with 0.114 moles of H_2O. Ratio these numbers to find x.	$0.114/0.0199 = 5.73$
x must be a whole number, so round correctly to the answer.	$x = 6$

Other gravimetric methods

Gravimetric methods are methods relating to masses or changing masses. There is a large variety of possible questions, but all rely on your knowledge of the chemistry involved and careful calculations.

EXAMPLE 6.2

A student adds an excess of silver nitrate solution, $AgNO_3(aq)$ to 50 cm^3 of a solution of potassium bromide of unknown concentration. The resultant mixture is filtered using a weighed filter paper, and the paper and residue is left to dry. The mass of the residue and paper is then recorded in a table.

Mass of dry filter paper/g	0.08
Mass of dry filter paper and residue/g	3.56

The precipitation reaction forms solid silver bromide, AgBr.

$$Ag^+(aq) + Br^-(aq) \rightarrow AgBr(s)$$

The AgBr(s) is the residue on the filter paper. Calculate the concentration of the potassium bromide solution.

Method

Calculate the mass of residue. This is the dry AgBr that was in the solution.	$3.56 - 0.08 = 3.48\,g$ of AgBr
Calculate the number of moles of AgBr. This is the number of moles of $Br^-(aq)$ and hence the number of moles of KBr(aq) in the original solution.	$3.48/(107.9 + 79.9) = 0.0185$ moles of KBr
This KBr was originally in 50 cm^3. Use this to find the original concentration.	$0.0185/(50/1000) = 0.371\,mol\,dm^{-3}$ of KBr

Determining an enthalpy change

Determining an enthalpy change involves using a reaction to change the temperature of a calorimeter. The calorimeter could be a volume of water or solution. From the temperature change and the heat capacity of the calorimeter, the energy released (or absorbed) by the reaction is measured. This is divided by the number of moles of reactant to give the ΔH value in $kJ\,mol^{-1}$ for the reaction.

Determining ΔH_c for a combustion reaction

The enthalpy change of combustion (ΔH_c) is usually determined for a liquid hydrocarbon or alcohol by burning it in a spirit burner and using the energy released to heat a volume of water. The apparatus used (Figure 6.1) is designed to allow the compound to burn while collecting as much heat as possible.

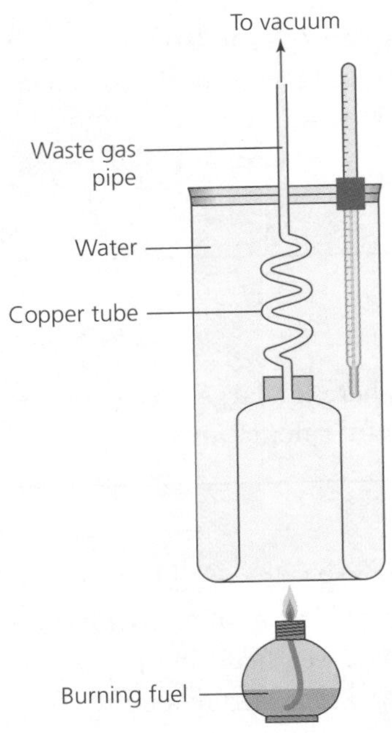

Figure 6.1 Equipment for determining enthalpy change of combustion

1. Measure the mass of the spirit burner.
2. Measure the initial temperature of the water.
3. Light the spirit burner and place it beneath the apparatus.
4. Stir the water continuously while monitoring the temperature rise.
5. When the temperature rise is approximately 50 °C, extinguish the spirit burner but keep stirring the water.
6. Record the highest temperature reached by the water.
7. Measure and record the final mass of the spirit burner.

The specific heat capacity of water is $4.18\,J\,g^{-1}\,K^{-1}$ and the energy required to heat the water, q, can be calculated from the equation $q = mc\Delta T$ (Figure 6.2). The change in temperature, ΔT, should be calculated as (initial temperature – final temperature). This means a temperature increase results in a negative value for ΔT, which will result in a negative final value for ΔH for the exothermic reaction.

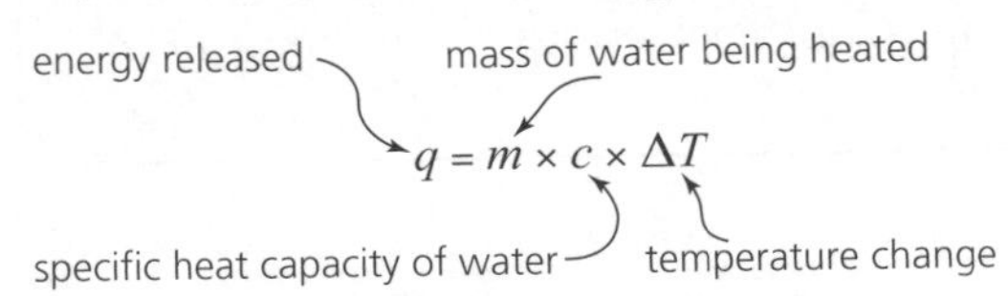

Figure 6.2 Equation for enthalpy calculation

The difference in mass measurements for the spirit burner is the mass of fuel combusted. This can be divided by the molecular mass of the fuel to give the number of moles combusted. Dividing q by the number of moles combusted gives ΔH_c.

EXAMPLE 6.3

A student conducts an experiment to determine ΔH_c for ethanol, C_2H_5OH. They used a calorimeter containing 200 cm^3 of water and recorded the results in the table.

Initial mass of spirit burner containing ethanol/g	62.46
Final mass of spirit burner containing ethanol/g	60.33
Initial temperature/ °C	21.5
Final temperature/ °C	73.0

Method

Calculate the temperature change, ΔT — $21.5 - 73.0 = -51.5\,°C$

Calculate the energy change for the reaction, q, using $q = mc\Delta T$ — $200 \times 4.18 \times -51.5 = -43\,054\,J$

Calculate the mass of C_2H_5OH combusted — $62.46 - 60.33 = 2.13\,g$

Calculate the moles of C_2H_5OH combusted (the molecular mass is $46\,g\,mol^{-1}$) — $2.13/46 = 0.0463$ moles

Calculate ΔH_c for ethanol by dividing the energy value by the number of moles combusted — $-43\,054/0.0463 = -929\,892\,J\,mol^{-1}$

Divide your answer by 1000 to give the more common units, $kJ\,mol^{-1}$ — $-929\,892/1000 = -930\,kJ\,mol^{-1}$

The actual value for ΔH_c (C_2H_5OH) is closer to $-1370\,kJ\,mol^{-1}$, so this experimental technique is not very accurate. The main reasons for this are:

- Heat loss to the surroundings – the waste gases carry much of the energy away from the reaction and do not transfer it efficiently to the water being heated.
- The reaction may be incomplete – spirit burners are not very efficient at burning liquid fuels and the incomplete combustion reaction produces carbon deposits and less heat.
- If the spirit burner is left uncovered while unlit, it will constantly lose mass by evaporation of fuel.

The method can be made much more accurate by combusting a fuel with a known ΔH_c, and using the data to calculate a better value for mc in the equation $q = mc\Delta T$. This is effectively calibrating the apparatus by calculating its heat capacity, rather than approximating it to the heat capacity of just the water being heated.

Determining ΔH_r for a reaction in solution

It is possible to get quite accurate measurements of ΔH_r for reactions that happen in solution by conducting the experiment in a polystyrene cup (Figure 6.3). The energy released changes the temperature of the solution and the value of q can be calculated using $q = mc\Delta T$ by approximating the specific heat capacity of the solution to that of water. The value of ΔH_r is then calculated as for ΔH_c above.

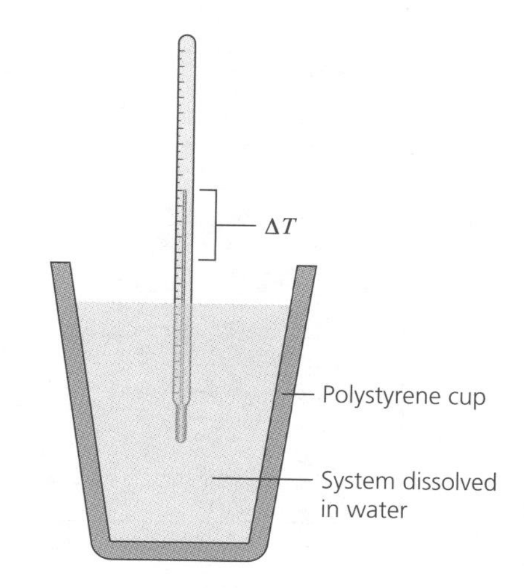

Figure 6.3 Polystyrene cup calorimeter

As soon as the temperature rises above room temperature, some heat will be lost to the surroundings despite the insulation of the cup. This makes the value for ΔT lower than it should be. To account for this, temperature measurements are recorded throughout the experiment. First, several measurements are made to establish a 'base line' initial temperature. Then the reactants are mixed and the value for ΔT is found by recording the temperature for several minutes and drawing a graph of the results. The graph in Figure 6.4 shows a cooling curve which can be extrapolated back to find the original ΔT.

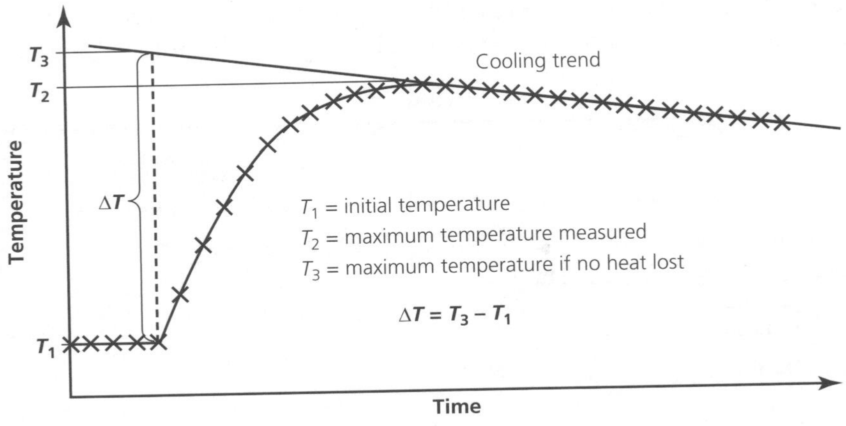

Figure 6.4 Cooling curve

EXAMPLE 6.4

A student adds 25.0 cm^3 of 2.00 mol dm^{-3} HCl to 25.0 cm^3 of 2.00 mol dm^{-3} NaOH in a polystyrene cup and takes the measurements shown in the table. Calculate ΔH_r for the neutralisation.

Time /min	Temperature /°C	Time /min	Temperature /°C	Time /min	Temperature /°C
0.0	21.0	3.0	30.5	6.0	31.0
0.5	21.0	3.5	31.5	6.5	30.5
1.0	21.0	4.0	32.0	7.0	30.0
1.5	21.0	4.5	32.0	7.5	30.0
2.0*	21.0	5.0	31.5	8.0	29.5
2.5	25.0	5.5	31.5	8.5	29.0

*The HCl(aq) is added at 2 minutes.

Method

Plot a graph of temperature against time. Draw a line of best fit for the temperatures up to 2 minutes, then a second line of best fit for the cooling section of the graph. ΔT is then found as shown in Figure 6.5.

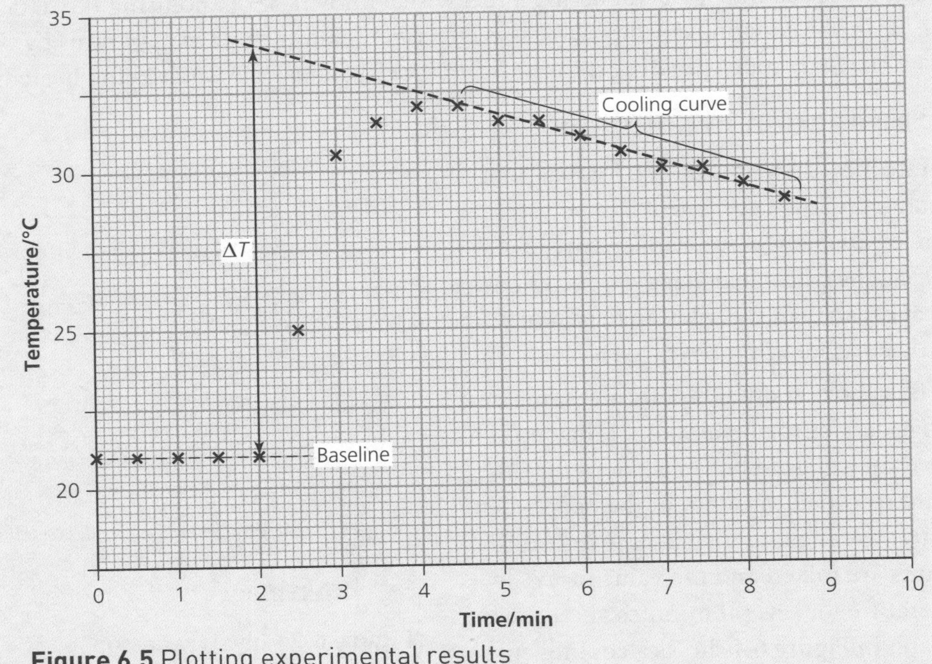

Figure 6.5 Plotting experimental results

Calculate the temperature change, ΔT	$21.0 - 34.0 = -13.0\,°C$
Calculate the energy change for the reaction, q, using $q = mc\Delta T$. Remember the volume of the solution is $50\,cm^3$ after the addition of the HCl	$50 \times 4.18 \times -13.0 = -2717\,J$
Calculate the moles of HCl reacted by using the concentration and initial volume	$2.00 \times (25/1000) = 0.0500$ moles
Calculate ΔH_r for 1 mole of HCl reacting	$-2717/0.0500 = -54\,340\,J\,mol^{-1}$
Divide your answer by 1000 to give the more common units, $kJ\,mol^{-1}$	$-54\,340/1000 = -54.3\,kJ\,mol^{-1}$

Finding a rate of reaction

Rates of reaction recorded at different concentrations of reactant are used to determine orders of reaction and hence rate constants.

The rate can be found by plotting the amount of a reactant or product (mass, volume or concentration) against time and measuring the gradient (Figure 6.6).

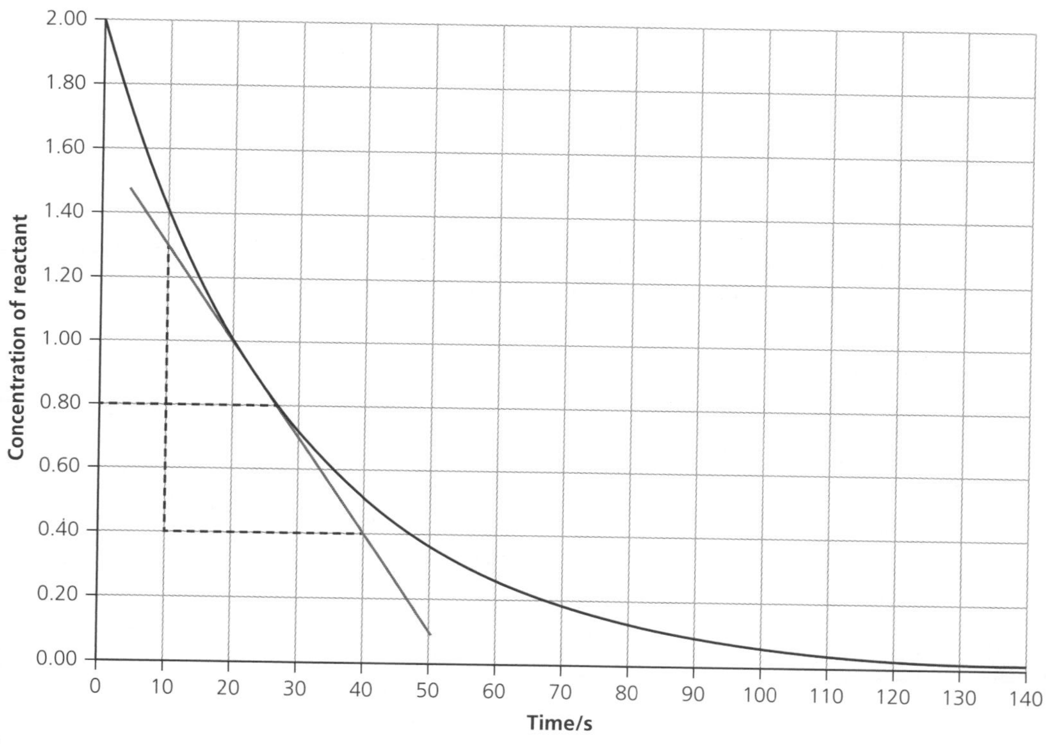

Figure 6.6 Concentration/time graph

The rate is found by drawing a tangent to the curve at the required concentration, and measuring the gradient of the tangent. The tangent should be a straight line that touches the curve only once, at the desired concentration. In the example, the rate at [reactant] = $0.80\,mol\,dm^{-3}$ is given by the rise/run:

$$\text{rate} = \frac{1.26 - 0.40}{40 - 10} = \frac{0.86}{30} = 0.029\,mol\,dm^{-3}\,s^{-1}$$

The relationship between reactant concentration and rate of reaction is normally found by plotting the rate against the concentration. The other reactant(s) are in excess so that their concentrations stay effectively constant and do not contribute to the change in rate.

The relationship between the concentration of a reactant and the rate is dependent on its order of reaction as shown in Figure 6.7.

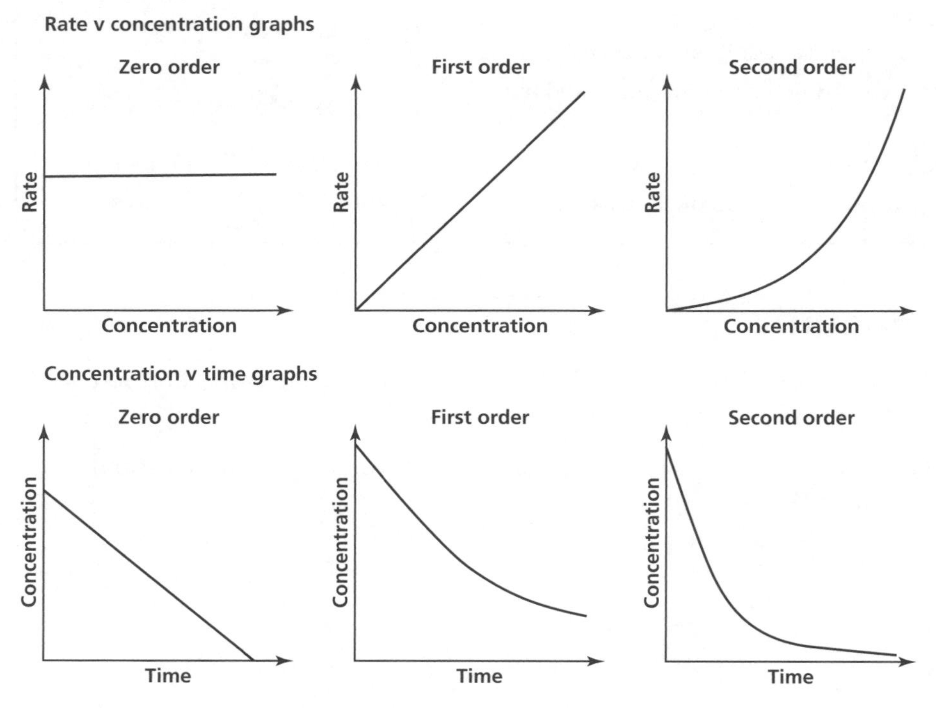

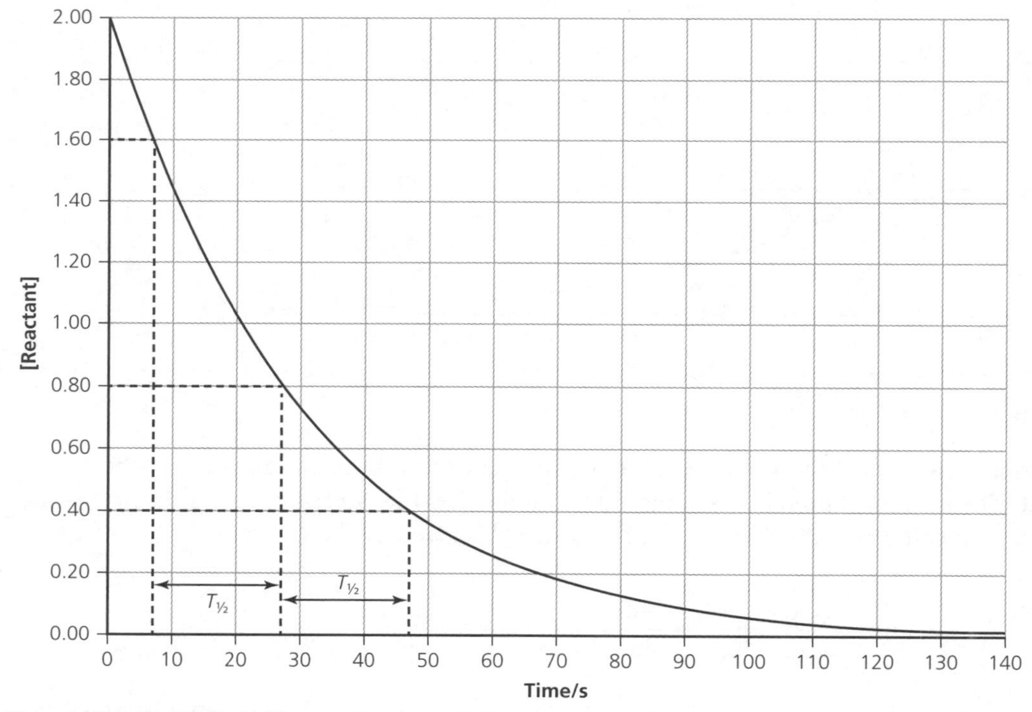

Figure 6.7 Shapes of reaction rate graphs

For first order reactions, the graph of concentration against time has a constant half-life, $T_{½}$. This means that the time taken to halve the concentration is constant regardless of where on the graph it is measured. It is often useful to draw 'construction lines' on the graph to show this, as in Figure 6.8.

Figure 6.8 Half-life in first order reactions

In this example, construction lines have been drawn at concentrations of 1.60, 0.80 and 0.40 mol dm^{-3}. The times given are approximately 7, 27 and 47 s, showing a constant half-life of 20 s.

For all rate against concentration graphs, ratios can be used to determine orders of reaction:

$$\frac{\text{rate}_1}{\text{rate}_2} = \frac{[\text{reactant}]_1^x}{[\text{reactant}]_2^x}$$

The ratio of rates at any two points on the graph is equal to the corresponding ratio of the reactant concentrations raised to their order of reaction (x).

Measuring a rate of reaction involves determining a changing physical property (for example, colour, gas volume or mass) that can be measured as the reaction proceeds.

Reactions with a colour change

Reactions that involve a colour change either produce an abrupt colour change (such as in the case of clock reactions) which can be timed and approximated to the rate, or a more gradual colour change which can be measured with colorimetry.

Clock reactions

Clock reactions only approximate the initial rate as the measurement occurs some time into the experiment, when the concentrations of reagents have changed. The reactions involved normally produce iodine, so a small amount of thiosulfate, $S_2O_3^{2-}$, is added (much smaller than the amount of iodine produced by the complete reaction) to convert the initial amount of iodine produced back to iodide.

$$I_2(aq) + 2S_2O_3^{2-} \rightarrow 2I^-(aq) + S_4O_6^{2-}(aq)$$

When the thiosulfate is used up, the iodine still being produced by the reaction being studied changes the colour of the solution. The effect is greatly enhanced by adding starch solution to the reaction mixture, which gives an abrupt change in colour from blue-black to colourless.

The method is useful to determine the orders of reaction as the ratio of 1/time taken for each reaction matches the ratio of rates of reaction at different reactant concentrations. In addition, as the thiosulfate converts the initial iodine produced back to iodide, the initial concentration of iodide can be considered constant.

This method requires a series of reactions to be conducted, to give rates for a series of concentrations of each reagent. This allows rate against concentration graphs to be drawn and the orders determined. When designing these experiments, if care is taken to keep the overall volume constant, the ratio of reagent volumes matches the ratio of concentrations. For example, in the following reaction:

$$IO_3^-(aq) + 5I^-(aq) + 6H^+(aq) \rightarrow 3I_2(aq) + 3H_2O(l)$$

the combinations of reagents shown in Table 6.1 could be used.

Experiment	Volume of IO_3^-/cm^3	Volume of I^-/cm^3	Volume of H^+/cm^3	Volume of $S_2O_3^{2-}$/cm^3	Volume of H_2O/cm^3
1	2	2	2	5	4
2	4	2	2	5	2
3	6	2	2	5	0
4	2	4	2	5	2
5	2	6	2	5	0
6	2	2	4	5	2
7	2	2	6	5	0

Table 6.1

- Experiments 1, 2 and 3 change only the concentration of IO_3^-.
- Experiments 1, 4 and 5 change only the concentration of I^-.
- Experiments 1, 6 and 7 change only the concentration of H^+.

The results of the first experiment are used in all three order determinations. Water is added to some experiments to keep the total volume at 15 cm^3 for all the experiments. The volume of $S_2O_3^{2-}$ is the same throughout, so that the rates are comparable.

When considering the method for the experiment, the reagents should all be added in the same way. Normally you put all of the reagents except one into a boiling tube and mix thoroughly. You then add the final reagent to start the reaction and start the clock. The clock is stopped when the colour change is seen and the time recorded. The rate of reaction is proportional to 1/time (1/*t*), and these rates can be compared to determine the orders of reaction for each reactant.

Using a colorimeter

Colorimeters pass light of a set colour (or wavelength) through a sample and record the absorbance of the light. The sample is normally a solution held in a cuvette, a small plastic box with an open top and opposite sides transparent to visible light.

As the reaction proceeds (and the solution changes colour), the absorbance of the light passing through the solution changes and this indicates the concentration of the coloured reagent or product. This is a more useful technique than the clock reaction as it can be applied to many more reactions.

The relationship between the concentration of a coloured compound and the absorbance is assumed to be linear and therefore the absorbance is proportional to the concentration. It is important to zero the colorimeter with a cuvette containing just water before the experiment begins and to select the colour of the light to match the colour absorbed by the compound. For example, bromine appears orange in solution as it absorbs blue light and transmits the other wavelengths. So to study the concentration of aqueous bromine, the calorimeter is set to measure the absorbance of blue light.

Reactions that have gaseous products

There are two methods to follow the progress of reactions that produce gaseous products. You can either record the change in mass of the reacting mixture or collect the gas and record its increase in volume.

Changing mass

The reaction is normally done in a conical flask on top of a mass balance (Figure 6.9). As the reaction proceeds, the mass of the reaction mixture reduces as the gaseous product is generated and escapes through the cotton wool plug (the plug is to ensure no spitting liquid leaves the flask, only the gas). The mass lost is plotted against time to show the progress of the reaction.

This method works well with a high precision balance (three decimal places) and a high M_r gaseous product, such as carbon dioxide, CO_2.

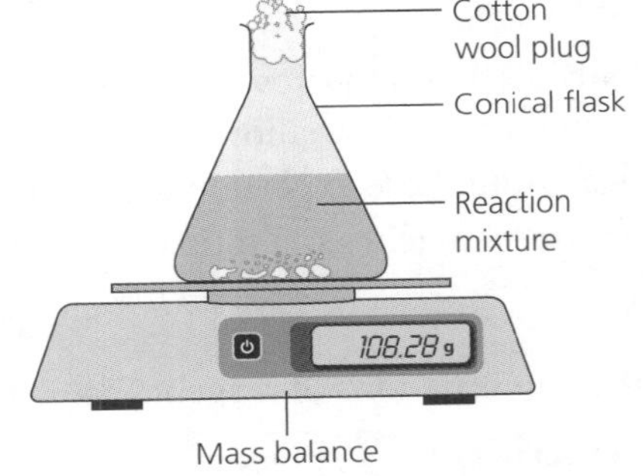

Figure 6.9 Measuring change of mass for reactions with gaseous products

EXAMPLE 6.5

A student reacts an excess of calcium carbonate, $CaCO_3$, with 50 cm^3 of 1.00 mol dm^{-3} sulfuric acid, H_2SO_4, and measures the mass at various times ($mass_t$) during the reaction. The final mass ($mass_f$) is recorded as the last reading. The value $mass_t - mass_f$ is proportional to the concentration of H_2SO_4 in the reactant mixture.

Calculate the value of $mass_t - mass_f$ for each measurement shown in the table and graph this value against time. Use your graph to show that the reaction is first order with respect to [H_2SO_4].

time/s	$mass_t$/g	$mass_t - mass_f$/g
0	73.02	
20	72.21	
40	71.70	
60	71.37	
80	71.17	
100	71.04	
120	70.96	

Method

First, complete the table. The value for $mass_f$ is the final mass recorded, 70.96 g.

time/s	$mass_t$/g	$mass_t - mass_f$/g
0	73.02	2.06
20	72.21	1.25
40	71.70	0.74
60	71.37	0.41
80	71.17	0.21
100	71.04	0.08
120	70.96	0.00

Then, graph $mass_t - mass_f$/g against time/s.

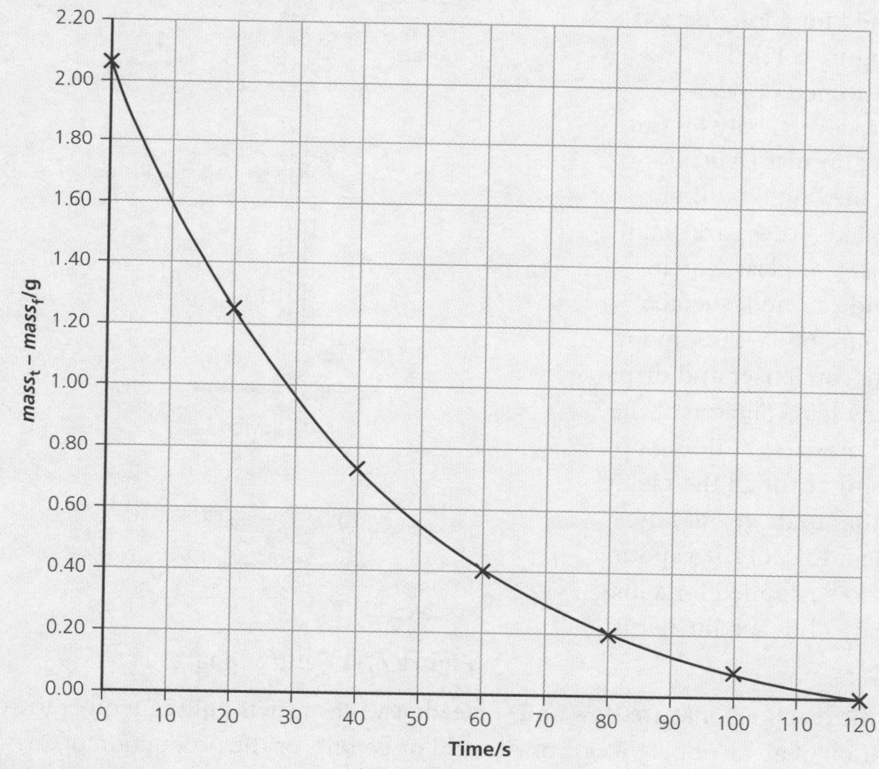

Figure 6.10

A line of best fit through the points can be drawn to give a smooth curve. Inspection of the graph shows that the value for $mass_t - mass_f$, proportional to $[H_2SO_4]$, halves approximately every 30 s. This constant half-life is a property of a first order reaction, so as the concentration of $CaCO_3$ is constant (as it is a solid), the order of reaction with respect to $[H_2SO_4]$ must be 1.

Measuring the volume of gas produced

There are two common ways in which this is done, either by using a gas syringe or by collecting the gas over water in an inverted measuring cylinder or burette (Figure 3.2).

In both cases, as the molar volume of any gas is constant at a given temperature, the volume of gas is proportional to the amount of gaseous product produced. Therefore a graph of volume against time is sufficient to record the progress of a reaction and monitor the rate.

Other methods

A final method to consider here is sampling. A reaction between two solutions can be monitored at various stages by titrating samples of the solution as the reaction proceeds. At certain time intervals, a sample of the reaction mixture is extracted (usually with a pipette) and transferred to a clean conical flask. The reaction in the conical flask is then 'quenched' (stopped), by adding a large excess of water or plunging into ice. The concentration of a reactant or product is then found by titration against the sample. In this way, the concentration of a reactant or product at various intervals during the reaction can be found, plotted and the rate found.

Organic preparations

Reactions involving organic reagents are often used to prepare samples of a pure organic compound and then test the purity of the product. The method used depends on the compound being prepared.

Reflux

When a reaction needs to be heated for a long period of time (hours), reflux is used (Figure 6.11). The apparatus consists of a round-bottomed or pear-shaped flask connected directly to a vertically mounted Liebig condenser. Anti-bumping granules (ABGs) are added to the flask to prevent the liquid boiling violently or 'bumping' as it is heated – these are small crystals of aluminium oxide, Al_2O_3, or glass which do not chemically react but provide a rough surface for bubbles to form on. As the liquid boils, the vapour produced condenses in the Liebig condenser and drips back into the flask, preventing any loss. The top of the condenser is left open to prevent pressure building up in the apparatus. The water passing through the Liebig condenser must enter at the bottom to drive out any air in the water jacket and enable it to cool the vapour effectively. As steady heat needs to be applied for a long time, a heating mantle rather than a Bunsen burner is used.

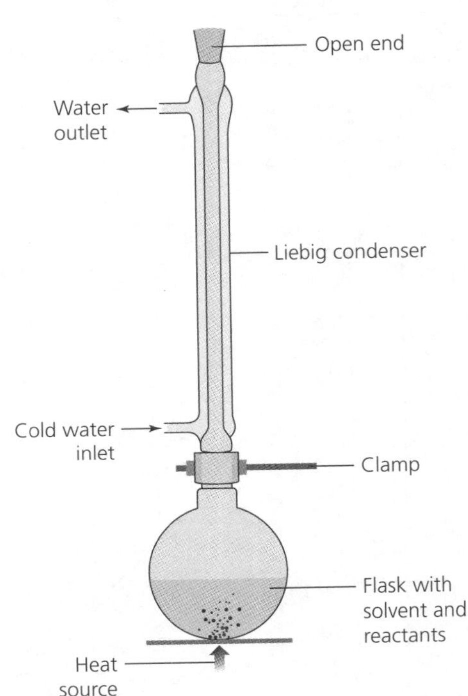

Figure 6.11 Reflux apparatus

This technique is useful for organic reactions that are slow and so need heating at their boiling temperature for a long time, such as the oxidation of an alcohol to a carboxylic acid or ketone, or the production of an ester from an alcohol and carboxylic acid (or indeed the hydrolysis of an ester).

A 'drying tube' may be added to the top of the apparatus to prevent water vapour from the air entering the reaction mixture. This consists of a glass tube packed with a desiccant such as calcium chloride, $CaCl_2$, which allows the passage of air but absorbs any water vapour.

Distillation

Distillation is a technique to separate two liquids by virtue of the difference in their boiling points. It can be used in organic preparations when a product has a significantly lower boiling point than the reactant mixture, such as the preparation of an aldehyde by oxidation of a primary alcohol.

The reaction mixture is heated with ABGs and the component with the lowest boiling point vaporises and moves through the apparatus until it meets the condenser, where it condenses and is collected in the collection vessel (Figure 6.12). The temperature of the vapour is measured with the bulb of the thermometer set in the T-piece which connects the round-bottomed flask to the condenser.

In practice, this technique serves only to concentrate the product rather than separate it completely from the other chemicals present, and further purification needs to be carried out (see following page).

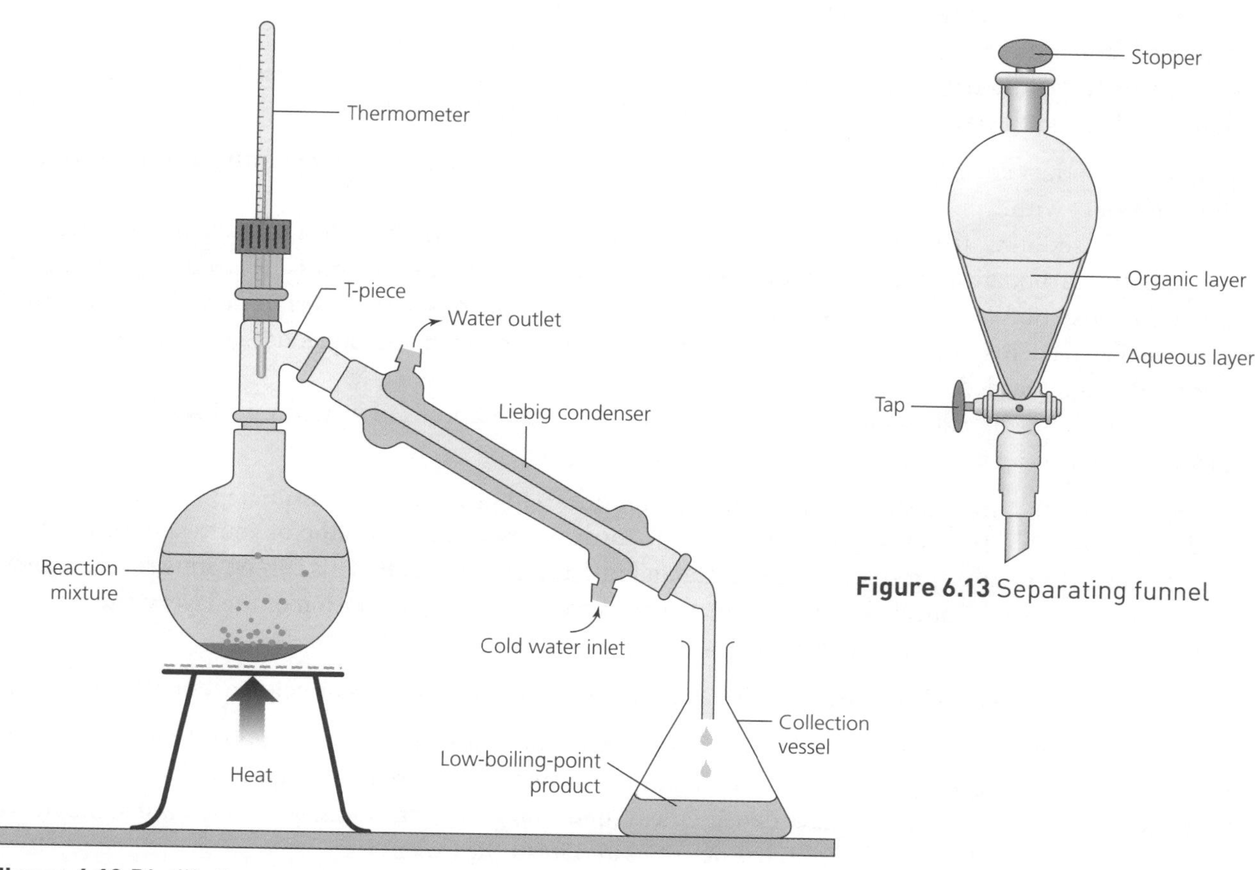

Figure 6.12 Distillation apparatus

Figure 6.13 Separating funnel

There are many variations to the distillation apparatus set-up:

- A **fractionating column** can be included between the round-bottomed flask and the T-piece. This is a glass tube with either glass beads or a glass coil inside it which makes the separation of the vapour more efficient as it gives a surface on which vapour of higher-boiling-point chemicals can condense and return to the reaction vessel before entering the Liebig condenser.
- For organic chemicals which decompose, it is useful to reduce their boiling points in some way. This can be achieved by sealing the apparatus and connecting it to a vacuum line to lower the pressure. This allows the distillation to occur at lower temperatures and can allow organic liquids to boil without decomposing. This technique is called **vacuum distillation**.
- As long as it will not react with the organic reagents, water can be added to the reaction mixture to reduce the temperature. This limits the boiling temperature to about 100 °C. The flow of steam through the apparatus takes the low-boiling-point organic product with it to the collection vessel, where it forms an organic layer which can be separated from the water. This technique is called **steam distillation**.

Purifying a liquid organic product

Liquid organic products are purified by distillation and then by the use of a separating funnel to remove aqueous impurities, such as acids.

A separating funnel is a simple piece of glassware with a tap at one end, a stopper at the other and a bulb of glass in between. It is used with a stand which holds the separating funnel upright with the tap at the bottom. They are used to allow two immiscible liquids to meet and be shaken together, then allowed to separate. The layers can be drawn off separately.

The impure organic product to be purified is added to the separating funnel along with another chemical which will abstract (dissolve into itself) either the desired product or the impurities. Normally the second layer will be aqueous, but this is not always the case.

The stopper is added and the two layers shaken together, with the pressure being released regularly by inverting the apparatus and briefly opening the tap before shaking again. The separating funnel is then placed into its stand and left while the layers separate. When two layers are clearly seen, the lower layer can be removed through the tap leaving the upper layer in the separating funnel.

To purify the product collected by distillation from an acid-catalysed reaction, a separating funnel is used to wash the product with:

- a solution of sodium carbonate, Na_2CO_3, which will form an aqueous layer. Be aware that the reaction between the carbonate and the acid will produce carbon dioxide gas, so the pressure needs to be released often. This reaction neutralises any acid in the product. The aqueous layer is then removed and discarded.
- deionised water. This dissolves any water-soluble impurities and forms an aqueous layer that can be removed and discarded.
- saturated sodium chloride solution, NaCl. This removes large amounts of water impurities from the product. It forms an aqueous layer that can be discarded.

To extract the product from a steam distillation (the product will be heavily contaminated with water):

- add an organic solvent, such as ether or hexane, to form an organic layer on top of your water product. On shaking, the organic product will be abstracted into the organic layer. This can be done several times to the same sample to collect as much product as possible. Following separation of the layers, the product can be recovered by evaporation of the solvent.

Care needs to be taken to release the pressure, especially for volatile solvents or acid/carbonate mixtures.

Care should also be taken to be aware of which layer holds your desired product; the upper layer is normally the organic layer and the lower the aqueous layer, but not always. The more dense liquid will form the bottom layer. Most organic solvents have densities less than 1.00 $g\,cm^{-3}$, the density of water, but some such as dichloromethane (1.33 $g\,cm^{-3}$) have higher densities and form the lower layer.

After exposure to water in a separating funnel, the final stage is to dry the product using a drying agent such as calcium chloride, $CaCl_2$. A small amount of anhydrous $CaCl_2$ is added to the product in a beaker or conical flask and they are swirled together. The $CaCl_2$ absorbs the final traces of water and the pure product can then be decanted (poured off the top) into a clean sample container.

Purifying a solid organic product

Solid organic products are purified by recrystallisation. This is done by identifying a solvent in which the pure product is insoluble at low temperatures but soluble at high temperatures.

The minimum amount of hot solvent is added to dissolve the impure product. Any insoluble impurities remain solid, so a hot filtration (where the filter funnel and paper are heated) allows the solid impurities to be separated.

The mixture is then allowed to cool, and crystals of the pure product form. As the crystals grow slowly, they exclude molecules of solvent and the soluble impurities. The pure solid is then separated by vacuum filtration (see below), washed with cold solvent and allowed to dry.

Vacuum filtration

This procedure separates solids from liquids much faster than normal filtration. It uses a Büchner funnel, a special filter funnel which includes a mesh to support a disc of filter paper (Figure 2.7, see page 10). The Büchner flask is a thick-walled side-arm conical flask which can support a vacuum without breaking. The reduced pressure pulls the solvent through the filter paper leaving the solid product behind.

Testing the purity of an organic product

It is important to determine the purity of the organic product that has been made. Common methods for doing this are **thin-layer chromatography** (TLC) and **melting point analysis**.

TLC is a technique that is used to separate mixtures for analysis, such as a product and impurities. A small amount of solid product is placed on a glass plate covered in silica, forming a spot. The plate is then mounted in a beaker with a small amount of solvent such that the solvent is able to run up the surface by capillary action. Different compounds will be carried different distances up the plate by the solvent. If the sample is pure, the spot moves up the plate in a uniform way and will not divide. If it is impure, more than one compound will be present and the spot will separate into two (or more). The final position of the compound can be seen using UV light, which causes the silica to fluoresce or glow.

The distance the spot moves from its starting location divided by the distance moved by the solvent is known as the R_F value (retention factor). The R_F value depends on the compound, solvent, temperature and other factors. It is possible to use this technique to identify a compound and this is usually done by putting a spot of pure, known compound next to the unknown one. If they move the same distance up the plate (and therefore have the same R_F value), they are likely the same compound.

A simpler technique to test the purity of a compound is to measure its melting point accurately. Pure compounds melt immediately at a specific temperature. Impurities normally lower the melting point temperature and cause the compound to melt over a temperature range. If the compound is pure, a carefully measured melting point can also be used to confirm its identity by reference to published data.

Additional safety considerations

An awareness of the chemical hazards involved with common reagents and products may be required and students should know how to minimise risks presented in the question.

Safety considerations in Table 6.2 are in addition to those listed at the start of the book (Chapter 1).

Risk	Response
Smelling chemicals	Always waft the vapour towards your nose rather than positioning your nose over the chemical
The experiment uses or generates dangerous (toxic, poisonous or corrosive) gases	Conduct the experiment in a fume cupboard
Boiling liquids in a test-tube	Use a (wider) boiling tube or heat gently. Do not point the open end of the tube at others
Boiling liquids in glassware	Use ABGs to prevent sudden boiling
The experiment uses chemicals that may catch fire	Keep the chemicals away from naked flames If the reaction needs to be heated to a temperature lower than 100 °C, use a water bath; otherwise use an electrical heater
Diluting concentrated sulfuric acid, a very exothermic process	Always add the acid slowly to water to prevent overheating

Table 6.2 Mitigating risks

A LEVEL

7 Practice questions

1 A student prepares a sample of butanone from butan-2-ol, both of which are soluble in water.

a State the reagent(s) needed to convert butan-2-ol to butanone. [1]

....

b The reaction is very slow at room temperature and needs to be heated. Suggest the best way to heat the reactants. Explain your answer. [2]

....

....

c The student plans to use reflux as the experimental technique. Draw a labelled diagram of the apparatus used for reflux. [4]

d The student starts by measuring 10.0 g of butan-2-ol, which is a liquid at room temperature. Describe how they can do this accurately. [2]

....

....

e Calculate the expected mass of butanone to be produced.

Expected mass of butanone = ... g [2]

f When the reaction is complete, the product is present in a mixture of water, acid and butanone. Describe how the student can obtain a sample of pure butanone from the mixture. [2]

....

....

....

....

g Following purification, the student found they had obtained 3.19 g of butanone.

i Calculate the percentage yield of butanone in this experiment.

Percentage yield of butanone = ... % [1]

ii Suggest a reason why the percentage yield is less than 100%. [1]

....

....

h Describe two chemical tests that, when used together, can confirm the presence of a ketone. Include the results of each test. [2]

....

....

....

....

[Total: 17]

2 A student wants to determine the concentration of a sample of potassium hydroxide, KOH. The pH of the sample was measured using a pH meter and found to be 14.52, but the pH meter is inaccurate.

The student transfers 20.0 cm^3 of the KOH solution into a glass beaker and places it beneath a burette filled with 2.50 mol dm^{-3} hydrochloric acid, HCl. The acid is then added 5 cm^3 at a time to the beaker, with the temperature measured after each addition.

a Write the equation for the reaction of potassium hydroxide with hydrochloric acid. [1]

....

....

The student tabulated their results below.

Vol of HCl added/cm^3	Temperature/°C
0.00	22.0
5.00	27.5
10.00	31.0
15.00	33.5
20.00	35.5

Vol of HCl added/cm^3	Temperature/°C
25.00	35.0
30.00	34.0
35.00	33.0
40.00	32.0
45.00	30.5

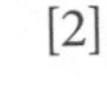

b Plot a graph of the vol of HCl added against temperature. [2]

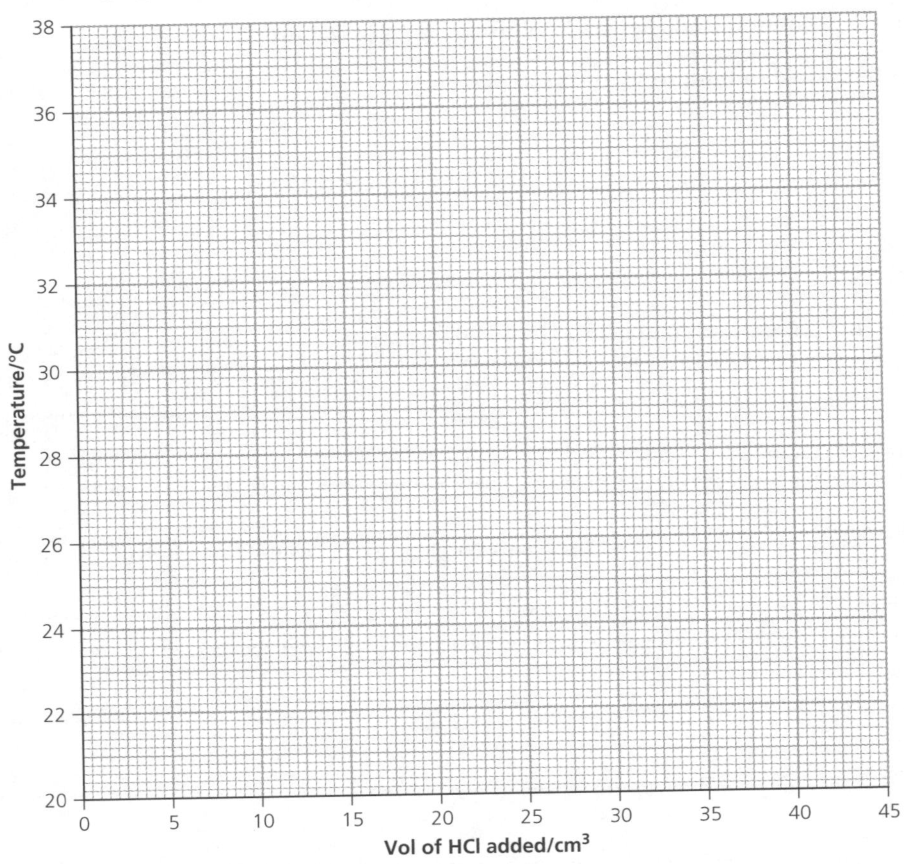

c Draw two lines of best fit on your graph, one for the heating curve (0–20 cm^3 added) and the other for the cooling curve (25–40 cm^3). Extrapolate both lines to find where they intersect. [2]

d From the intersect of the two lines, determine the volume of HCl that reacts exactly with 20.0 cm^3 of the sample of KOH.

The intersect is at .... cm^3 [1]

e Calculate the concentration of KOH in the sample.

Concentration KOH = ... mol dm^{-3} [1]

f Suggest three ways in which the experiment could be made more accurate. [3]

....

....

....

....

....

g The burette had previously been used in another experiment to measure the volume of a strong alkali. If the burette had not been rinsed with the acid for use in this experiment, state and explain the effect this would have had on the calculated value of the concentration of the KOH sample. [2]

....

....

....

....

[Total: 12]

3 Magnesium sulfate exists in a hydrated form which includes water of crystallisation and has the formula $MgSO_4.xH_2O$.

An experiment involving heating to a constant mass is conducted to determine the value of x.

a Explain what is meant by *heating to a constant mass* and state why it is used. [2]

....

....

....

b Write a brief method for the experiment used to determine x. You may assume you have all the apparatus normally found in a school laboratory and are provided with approximately 5 g of hydrated magnesium sulfate.

Use bullet points in your answer. [4]

....

....

....

....

....

....

....

c Identify two risks involved in the experiment and suggest ways to mitigate them. [2]

....

....

....

....

d The table gives results of another student's experiment to determine x.

Mass of crucible/g	48.22
Mass of crucible + sample/g	53.23
Mass of crucible + residue after first heating/g	51.59
Mass of crucible + residue after second heating/g	50.72
Mass of crucible + residue after third heating/g	50.71

i Use the student's results to calculate the value of x in $MgSO_4.xH_2O$. [4]

ii The student was concerned that the solid changed colour when heated strongly.

Suggest what could have happened and predict and explain the effect on the value of x calculated for $MgSO_4.xH_2O$. [3]

....

....

....

....

....

....

....

[Total: 15]

4 A student conducts an experiment to determine a value for the partition coefficient, K_{part}, between pentanoic acid, C_4H_9COOH, in cyclohexane, C_6H_{12}, and in water. Cyclohexane is a dangerous chemical and should not be used in practical activities.

The student begins with an aqueous solution of 0.400 mol dm^{-3} of C_4H_9COOH and the usual school laboratory apparatus. The student follows the procedure below.

- Prepare 100 cm^3 of a solution of 0.100 mol dm^{-3} $C_4H_9COOH(aq)$.
- Add 20 cm^3 of this solution and 20 cm^3 of C_6H_{12} into a separating funnel.
- Shake the liquids together well, then allow to fully separate.
- Run off the lower (aqueous) layer into a conical flask.
- Titrate the contents of the conical flask against 0.0100 mol dm^{-3} sodium hydroxide, NaOH.

At the end of the experiment, the student's titre was 7.80 cm^3.

a Suggest a risk involved in using cyclohexane and suggest a way to mitigate it. [1]

....

....

b Suggest a suitable indicator to use in the titration. Explain your answer. [1]

....

....

c Write the expression for K_{part} for pentanoic acid between cyclohexane and water.

K_{part} = [1]

d Several steps are needed to calculate the value of K_{part} for this system.

i Calculate the initial number of moles of C_4H_9COOH in the 20 cm^3 of 0.0100 mol dm^{-3} solution added to the separating funnel.

Initial moles C_4H_9COOH in the aqueous layer = ...
moles [1]

ii Calculate the number of moles of C_4H_9COOH in the final 20 cm^3 aqueous layer used in the titration.

Final moles of C_4H_9COOH in the aqueous layer = ...
moles [1]

iii Calculate the number of moles of C_4H_9COOH that transferred into the cyclohexane layer on shaking the two solutions together in the separating funnel.

Moles of C_4H_9COOH in the organic layer = ... moles [1]

iv Suggest why it is not necessary to calculate the concentrations of C_4H_9COOH in each layer, even though they are in the K_{part} equation. [1]

....

....

v Calculate the value of K_{part} for this system.

K_{part} = ... [1]

e When running off the aqueous layer from the separating funnel, some of the aqueous layer was left in the tap. State and explain the effect of this on the calculated value of K_{part}. [2]

....

....

....

....

[Total: 10]

5 A student plans to prepare a sample of pentanal, C_4H_9CHO. They have access to the usual school laboratory apparatus and the following chemicals:

- pentan-1-ol, $C_5H_{11}OH$
- concentrated sulfuric acid, H_2SO_4
- sodium dichromate, $Na_2Cr_2O_7$.

a During the reaction, the dichromate ion, $Cr_2O_7^{2-}$, is converted into the chromium(III) ion, Cr^{3+}.

Construct a half equation for this conversion in aqueous acid.

....

Use your half equation to explain how the dichromate ion is acting as an oxidising agent in the reaction. [2]

....

....

....

....

b Suggest the half equation for the conversion of pentan-1-ol to pentanal.

$$C_5H_{11}OH \rightarrow C_4H_9CHO + \text{..................} + 2e^-$$ [1]

c The reaction involves a colour change. In the table, state the colour change and identify the species responsible for it. [2]

	From	To
Colour		
Species responsible for colour		

d The boiling points for some compounds containing five carbon atoms and oxygen are shown in the table.

Substance	Boiling point/°C
pentan-1-ol, $C_5H_{11}OH$	137
pentanal, C_4H_9CHO	102
pentanoic acid, C_4H_9COOH	187

i Describe and explain the trend in boiling points shown in the table. [2]

....

....

....

....

ii Suggest a suitable experimental method for the preparation of pentanal from pentan-1-ol and draw a labelled apparatus diagram. [4]

Name of method: ....

Apparatus diagram:

iii Pentanal is only very slightly soluble in water. Suggest a possible contaminant from the reaction and a method for removing it, assuming complete conversion of pentan-1-ol to pentanal. [4]

....

....

....

....

....

....

e State a chemical test and positive result for the aldehyde group present in pentanal. [2]

....

....

....

....

f Methylpropanal, $CH_3CH(CH_3)CHO$, is another aldehyde. Suggest the features you would expect to see in the ^1H and ^{13}C spectra of methylpropanal. [5]

^1H spectra:

....

....

^{13}C spectra:

....

....

[Total: 22]

6 Hydrogen iodide, HI, is a colourless gas. Under certain conditions, it decomposes to form hydrogen and iodine.

$$HI + HI \rightarrow I_2 + H_2$$

An experiment was conducted to investigate the kinetics of the reaction.

0.080 moles of HI was added to a 1.00 dm^3 vessel at a temperature of 900 °C. The concentration of HI was measured over the first 20 s of the reaction.

Time/s	[HI]/mol dm^{-3}
0	0.0800
4	0.0311
8	0.0202
12	0.0151
16	0.0121
20	0.0101

a Suggest how the progress of the reaction could be followed. [2]

....

....

....

....

b Plot a graph of [HI] against time and draw a smooth curve through the points. [2]

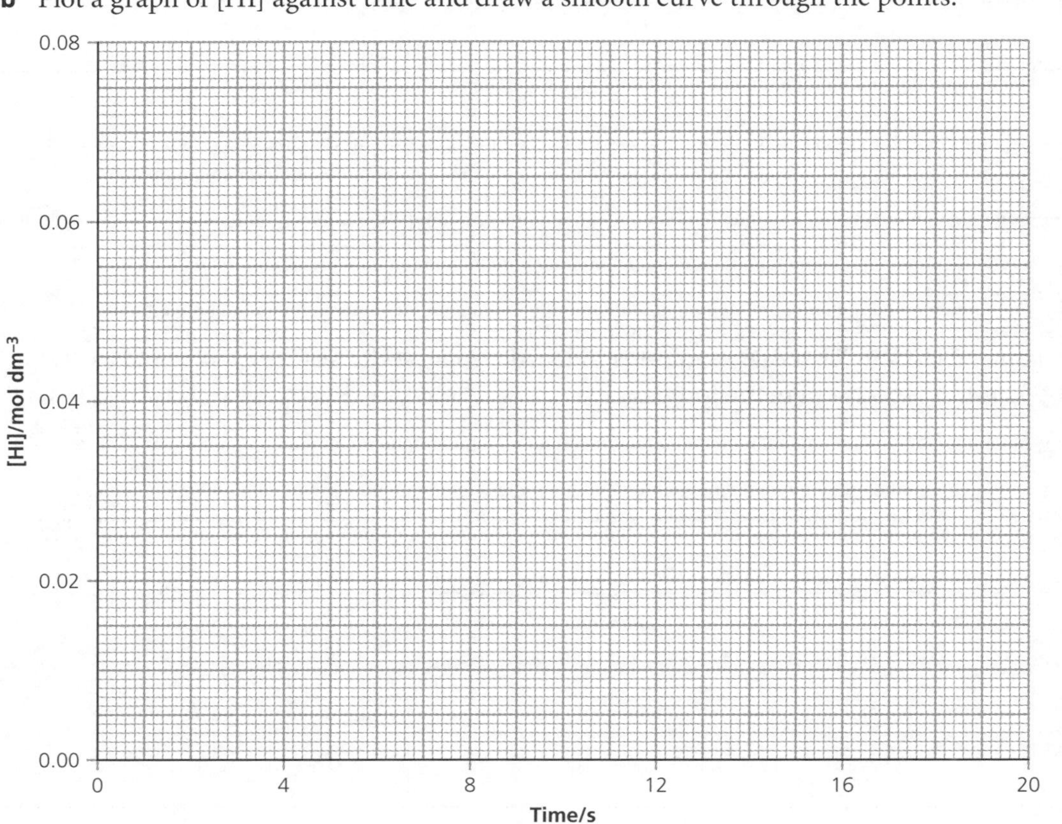

c Draw tangents to the curve at concentrations of [HI] = 0.0500 mol dm^{-3} and 0.0200 mol dm^{-3} and measure their gradients.

Give the co-ordinates of the two points you used for each of the tangents.

Tangent at [HI] = 0.0500 mol dm^{-3}

Co-ordinates .... Gradient ....

Tangent at [HI] = 0.0200 mol dm^{-3}

Co-ordinates .... Gradient .... [2]

d The gradients are proportional to the rate of reaction. Use the two gradients to calculate the order of reaction with respect to [HI]. [2]

e Deduce the rate equation and state the units for the rate constant, k.

Rate = ...

Units for k .... [2]

f A series of experiments was conducted at different temperatures to determine the activation energy, E_A, for the reaction. At each temperature, the rate constant was calculated.

Temperature, T (K)	k (units left blank)	$1/T$ (K^{-1})	$\ln(k)$
875	1.76	1.14×10^{-3}	0.565
900	3.87		
950	16.5		
1000	87.3		
1050	198		

i Complete the table by calculating $1/T$ and $\ln(k)$ for each temperature. [2]

ii Plot a graph of $\ln(k)$ against $1/T$.
Draw a line of best fit and determine the gradient. [3]

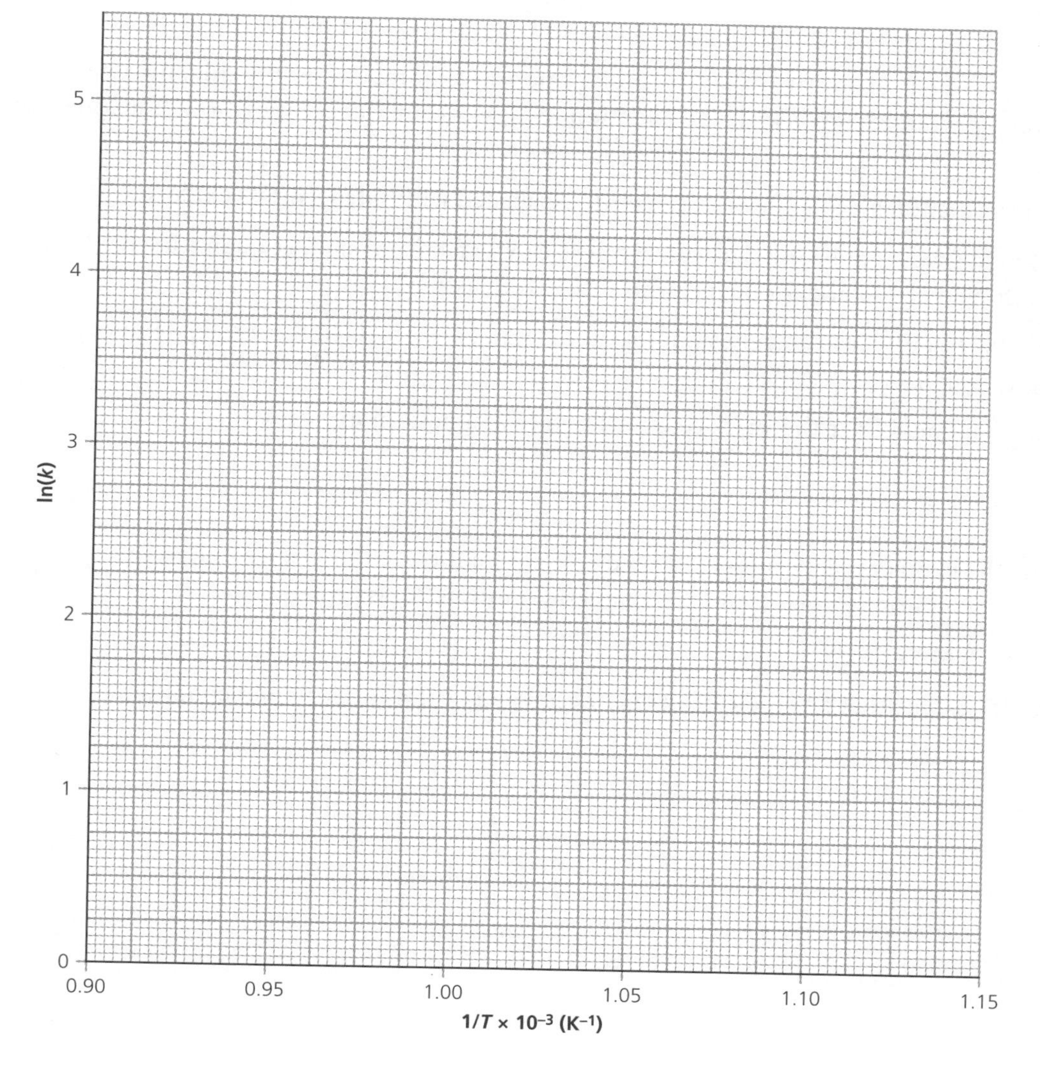

g From your gradient, determine the activation energy for the reaction.

Activation energy = ... $kJ\,mol^{-1}$ [3]

[Total: 18]

7 When magnesium nitrate(V) is heated, it decomposes to form magnesium oxide, nitrogen(IV) oxide and oxygen.

Nitrogen(IV) oxide is an acidic gas that reacts readily and completely with alkalis.

Plan a **single** experiment to confirm that the molar quantities of magnesium oxide, nitrogen(IV) oxide and oxygen produced agree with the equation for the thermal decomposition of magnesium nitrate(V).

The following information gives some of the hazards associated with nitrogen(IV) oxide.

Nitrogen(IV) oxide must not be inhaled. A large dose can be fatal and smaller quantities can have severe effects on breathing, particularly for people who suffer from asthma.

You are provided with anhydrous magnesium nitrate(V) and have access to the usual laboratory equipment and reagents.

a i Write an equation for the thermal decomposition of magnesium nitrate(V). [1]

....

ii Calculate the mass of magnesium oxide and volumes of nitrogen(IV) oxide and oxygen produced under room conditions when 1 mole of magnesium nitrate(V) is heated.

[A_r: O = 16.0; Mg = 24.3]

You should assume that one mole of any gas occupies $24.0\,dm^3$ under room conditions. [1]

b i Draw and label a diagram of the apparatus and experimental set-up you would use.

The set-up needs to be capable of absorbing the nitrogen(IV) oxide and collecting the oxygen separately and in sequence. [4]

ii State the volume of the gas collector to be used to collect oxygen in **i**. Calculate a mass of magnesium nitrate(V) to be heated that would produce a stated volume of oxygen appropriate for the collector.

[A_r: N, 14.0; O, 16.0; Mg, 24.3]

You should assume that one mole of any gas occupies 24.0 dm^3 under room conditions. [1]

c List the measurements you would make when carrying out the experiment. [3]

....

....

....

....

d i How could you make sure that the magnesium nitrate(V) had completely decomposed in the experiment? [1]

....

....

ii To make sure that the volume of gas measured is accurate, what should you do before taking the measurement? [1]

....

....

e Explain how you would use the results of the experiment to confirm that the decomposition had occurred according to the molar ratios in the equation. [2]

....

....

....

....

f What precautions would you take to make sure that the experiment could be performed safely? [1]

....

....

....

[Total: 15]

Cambridge International AS & A Level Chemistry 9701 Paper 51 Q1 June 2014

8 A saturated aqueous solution of magnesium methanoate, $Mg(HCOO)_2$, has a solubility of approximately 150 g dm^{-3} at room temperature. Its exact solubility can be determined by titrating magnesium methanoate against aqueous potassium manganate(VII).

During the titration, the methanoate ion, $HCOO^-$, is oxidised to carbon dioxide while the manganate(VII) ion, MnO_4^-, is reduced to Mn^{2+}.

You are supplied with:

- a saturated aqueous solution of $Mg(HCOO)_2$
- aqueous potassium manganate(VII), $KMnO_4$, of concentration 0.0200 mol dm^{-3}.

a **i** Write the half equations for the oxidation of $HCOO^-(aq)$ to $CO_2(g)$ and the reduction of $MnO_4^-(aq)$ to $Mn^{2+}(aq)$ in acid solution. [2]

....

....

ii Using the approximate solubility above, calculate the concentration, in mol dm^{-3}, of the saturated aqueous magnesium methanoate and the concentration of the methanoate ions present in this solution.

[A_r: H, 1.0; C, 12.0; O, 16.0; Mg, 24.3] [2]

iii In order to obtain a reliable titre value, the saturated solution of magnesium methanoate needs to be diluted.

Describe how you would accurately measure a $5.0\,cm^3$ sample of saturated magnesium methanoate solution and use it to prepare a solution fifty times more dilute than the saturated solution. [2]

....

....

....

....

iv Before the titration is carried out, dilute sulfuric acid must be added to the magnesium methanoate.

Explain why this is necessary and also whether the volume of sulfuric acid chosen will affect the result of the titration. [2]

....

....

....

....

v The potassium manganate(VII) is added from a burette into the magnesium methanoate in a conical flask.

Describe what you would see when you had reached the end-point of the titration. [1]

....

....

vi 1 mol of acidified MnO_4^- ions reacts with 2.5 mol of $HCOO^-$ ions.

$25.0\,cm^3$ of the diluted solution prepared in **iii** required $25.50\,cm^3$ of $0.0200\,mol\,dm^{-3}$ potassium manganate(VII) solution to reach the end-point.

Use this information to calculate the concentration, in $mol\,dm^{-3}$, of $HCOO^-$ ions in the diluted solution.

.... $mol\,dm^{-3}$ [1]

vii Use your answer to **vi** to calculate the concentration, in mol dm^{-3}, of the saturated solution of magnesium methanoate, $Mg(HCOO)_2$. Give your answer to **three significant figures**.

.... mol dm^{-3} [1]

b The solubility of magnesium methanoate can be determined at higher temperatures using the same titration.

In an experiment to determine how the concentration of saturated magnesium methanoate varies with temperature, name the independent variable and the dependent variable. [1]

Independent variable..

Dependent variable..

c The solubility of magnesium methanoate increases with temperature.

What does this tell you about ΔH for the process below?

$$Mg(HCOO)_2(s) \rightleftharpoons Mg^{2+}(aq) + 2HCOO^-(aq)$$

Explain your answer. [2]

....

....

....

....

d A student used the same titration method, this time to measure the concentration of a saturated solution of *barium* methanoate.

Explain why the acidification of the solution with dilute sulfuric acid might make the titration difficult to do. [1]

....

....

[Total: 15]

Cambridge International AS & A Level Chemistry 9701 Paper 51 Q1 June 2015

9 Propanone, CH_3COCH_3, is an organic liquid which is soluble in water.

Aqueous propanone reacts with aqueous iodine. The reaction is catalysed by $H^+(aq)$ ions.

$$CH_3COCH_3(aq) + I_2(aq) \rightarrow CH_3COCH_2I(aq) + HI(aq)$$

The order of reaction with respect to iodine can be determined experimentally.

An experiment is carried out using the following solutions:
- solution **A**, 25.0 cm^3 of 1.00 $mol\,dm^{-3}$ $CH_3COCH_3(aq)$
- solution **B**, 25.0 cm^3 of 1.00 $mol\,dm^{-3}$ $H_2SO_4(aq)$
- solution **C**, 50.0 cm^3 of 0.200 $mol\,dm^{-3}$ $I_2(aq)$.

The solutions are mixed to start the reaction. At certain time intervals, a 10.0 cm^3 portion of the mixture is withdrawn and transferred to a conical flask containing excess sodium hydrogencarbonate, $NaHCO_3(aq)$. This prevents any further significant reaction taking place by removing the H^+ (aq) ions. The concentration of unreacted $I_2(aq)$ in each 10.0 cm^3 portion of the mixture can then be determined by titration with aqueous thiosulfate ions, $S_2O_3^{2-}(aq)$.

a State the size and type of apparatus needed to prepare a suitable volume of a standard solution of 1.00 $mol\,dm^{-3}$ $CH_3COCH_3(aq)$ from liquid propanone.

Calculate the mass of propanone needed to prepare this standard solution.

[A_r: C, 12.0; H, 1.0; O, 16.0]

Apparatus..

Mass of propanone.. g [3]

b Solutions **A**, **B** and **C** need to be added in a specific order and the clock started as the third solution is added.

i Suggest the best order of adding the solutions.

1 ....

2 ....

3 .... [1]

ii Explain your choice. [1]

....

....

....

c Each 10.0 cm^3 portion of mixture removed from the main reaction is added to a separate solution of sodium hydrogencarbonate, $NaHCO_3(aq)$, in a conical flask to remove $H^+(aq)$ ions.

i Which piece of apparatus should be used to transfer each 10.0 cm^3 portion of mixture to the conical flask? [1]

....

ii Suggest **two** reasons why $NaHCO_3(aq)$ is preferred to $NaOH(aq)$ as the reagent used to remove $H^+(aq)$ ions.

Reason 1 ....

Reason 2 .... [2]

d The unreacted iodine in each 10.0 cm^3 portion of the mixture is titrated against 0.100 $mol\,dm^{-3}$ aqueous thiosulfate ions, $S_2O_3^{2-}(aq)$, to determine the concentration of $I_2(aq)$ in the mixture at the time that the 10.0 cm^3 portion was withdrawn.

$$I_2(aq) + 2S_2O_3^{2-}(aq) \rightarrow 2I^-(aq) + S_4O_6^{2-}(aq)$$

i A 10.0 cm^3 portion of mixture is removed at time = 0. This is before any of the 0.200 $mol\,dm^{-3}$ $I_2(aq)$ had reacted.

Calculate the volume of 0.100 $mol\,dm^{-3}$ $S_2O_3^{2-}(aq)$ needed to react with the iodine present in this 10.0 cm^3 portion of mixture.

Volume 0.100 $mol\,dm^{-3}$ $S_2O_3^{2-}(aq)$ = ... cm^3 [3]

ii Suggest the name of a suitable indicator to use in the titration and state its colour change.

Indicator ....

Colour change .... [2]

e State **two** variables which must be recorded in this experiment.

For each variable, state the units.

Variable 1 .... units ....

Variable 2 .... units .... [2]

f State **one** other variable which must be controlled in this experiment. [1]

....

....

g The order of reaction with respect to iodine is expected to be first order.

i Use the axes below to draw a sketch graph of how the concentration of iodine changes during the experiment. Label both axes. [2]

ii How could the graph be used to prove that the order of reaction with respect to iodine is first order? [1]

....

....

h A student suggested that the temperature at which the experiment was carried out would affect the order of reaction with respect to iodine.

State if the student was correct and explain your answer. [1]

....

....

[Total: 20]

Cambridge International AS & A Level Chemistry 9701 Paper 52 Q1 March 2016

10 The enthalpy change of reaction, ΔH_r, for the decomposition of sodium hydrogencarbonate, $NaHCO_3(s)$, cannot be measured directly.

$$2NaHCO_3(s) \rightarrow Na_2CO_3(s) + H_2O(l) + CO_2(g)$$

A student must carry out **two** separate experiments and use the results of these experiments to determine the enthalpy change of reaction for the decomposition of sodium hydrogencarbonate.

a Suggest why the enthalpy change of reaction, ΔH_r, for the decomposition of sodium hydrogencarbonate cannot be measured directly. [1]

....

....

In both experiments the student used a weighing boat. A weighing boat is a small vessel used to contain solid samples when they are weighed.

Experiment 1 Reaction between sodium carbonate, $Na_2CO_3(s)$, and dilute hydrochloric acid, HCl (aq)
step 1 The student added approximately 3 g of $Na_2CO_3(s)$ to a weighing boat and accurately measured the combined mass of the weighing boat and $Na_2CO_3(s)$. This mass was recorded in Table 7.1.
step 2 The student used a measuring cylinder to measure 50 cm^3 of 2 mol dm^{-3} HCl (aq).
step 3 The experiment was carried out and the results were recorded in Table 7.2.
step 4 The student reweighed the empty weighing boat and recorded the mass in Table 7.1.

Mass of weighing boat and $Na_2CO_3(s)$/g	4.15
Mass of empty weighing boat after addition of $Na_2CO_3(s)$ to HCl (aq)/g	0.97
Mass of $Na_2CO_3(s)$ added/g	

Table 7.1 Mass results from Experiment 1

Time/minutes	0	1	2	3		5	6	7	8	9	10
Temperature of mixture/°C	20.0	19.8	19.8	19.8		24.6	24.7	24.5	24.3	24.1	23.9

Table 7.2 Temperature results from Experiment 1

b i Outline how the student carried out **step 3** of the experiment.

You may find it helpful to write your answer as a series of smaller steps.

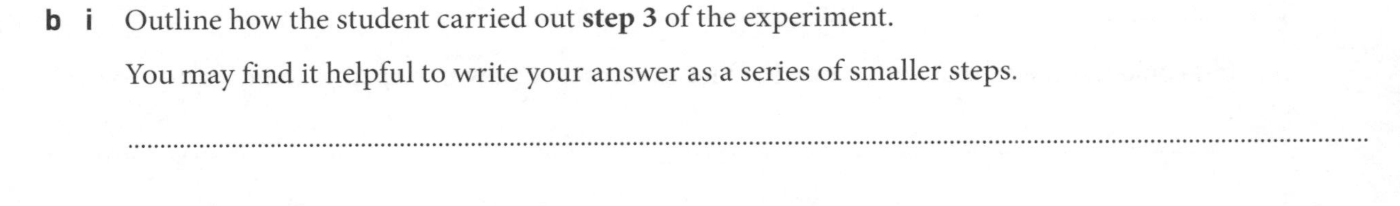

ii Draw a labelled diagram of the apparatus. [3]

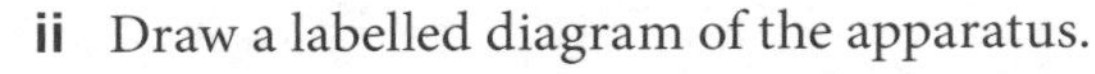

The student plotted a graph of the results and drew **two** lines of best fit which were both extrapolated as shown.

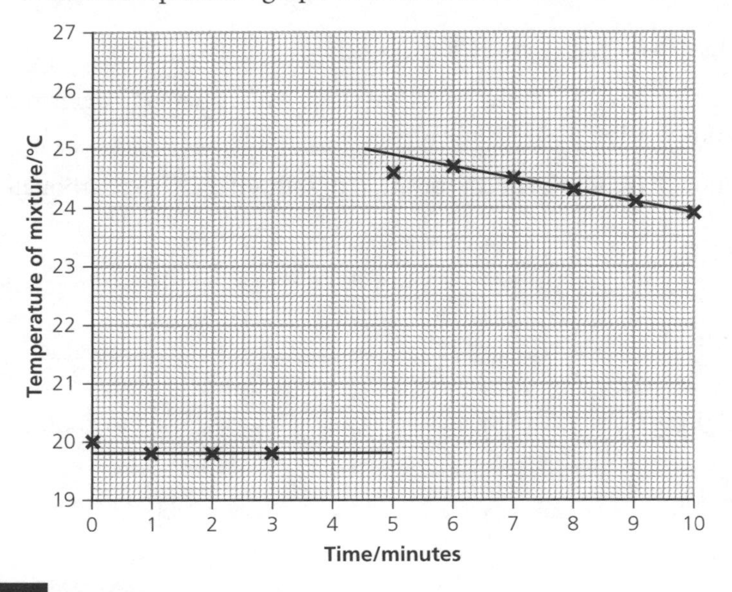

c Use the graph to determine the theoretical temperature increase at 4 minutes.

Theoretical temperature increase = ... °C [1]

d Use Table 7.1 on page 88 to determine the mass of $Na_2CO_3(s)$ which was added to the HCl (aq). Use this value and your answer to **c** to determine the enthalpy change, ΔH_1, for the reaction shown.

$$Na_2CO_3(s) + 2HCl(aq) \rightarrow 2NaCl(aq) + H_2O(l) + CO_2(g)$$

Give your answer to **three** significant figures.

(Assume that the specific heat capacity of the solution is $4.18\,J\,g^{-1}\,K^{-1}$.)

[A_r: Na, 23.0; C, 12.0; O, 16.0]

ΔH_1 = ... $kJ\,mol^{-1}$ [3]

e **i** Explain why the student did **not** add the $Na_2CO_3(s)$ to the HCl (aq) at 0 minutes. [1]

....

....

ii Suggest why the temperature measured at 5 minutes was lower than the temperature measured at 6 minutes. [1]

....

....

Experiment 2 Reaction between sodium hydrogencarbonate, $NaHCO_3(s)$, and dilute hydrochloric acid, HCl(aq)

- **step 1** The student weighed an empty weighing boat and recorded the mass in Table 7.3.
- **step 2** The student added exactly 4.20 g of $NaHCO_3(s)$ to the weighing boat and recorded the mass in Table 7.3.
- **step 3** The student carried out the same experimental procedure as in **steps 2** and **3** of Experiment 1.

Mass of empty weighing boat/g	0.95
Mass of weighing boat and $NaHCO_3(s)$/g	5.15
Mass of $NaHCO_3(s)$ added/g	

Table 7.3 Mass results from Experiment 2

f Explain why the method of determining the mass of solid added in Experiment 2 is less accurate than the method of determining the mass of solid added in Experiment 1. [1]

....

....

g **i** In Experiment 2 a 50 cm^3 measuring cylinder was used to measure the 50 cm^3 of HCl(aq). The 50 cm^3 measuring cylinder had 1 cm^3 graduations.

Calculate the maximum percentage error in measuring 50 cm^3 of HCl (aq) with this 50 cm^3 measuring cylinder.

Maximum percentage error = ... % [1]

ii Explain why measuring the concentration of the 2 mol dm^{-3} HCl more precisely would **not** affect the result of the experiment. [1]

....

iii Suggest what the student should change to reduce the percentage error associated with the temperature readings **without** changing the apparatus. [1]

....

....

h The student used the results from Experiment 2 and correctly determined the enthalpy change for the reaction between $NaHCO_3(s)$ and HCl(aq), ΔH_2, to be +24.2 kJ mol^{-1}.

$$NaHCO_3(s) + HCl(aq) \rightarrow NaCl(aq) + H_2O(l) + CO_2(g) \qquad \Delta H_2 = +24.2\ kJ\ mol^{-1}$$

Use the axes to draw a sketch graph of the expected results of Experiment 2. [2]

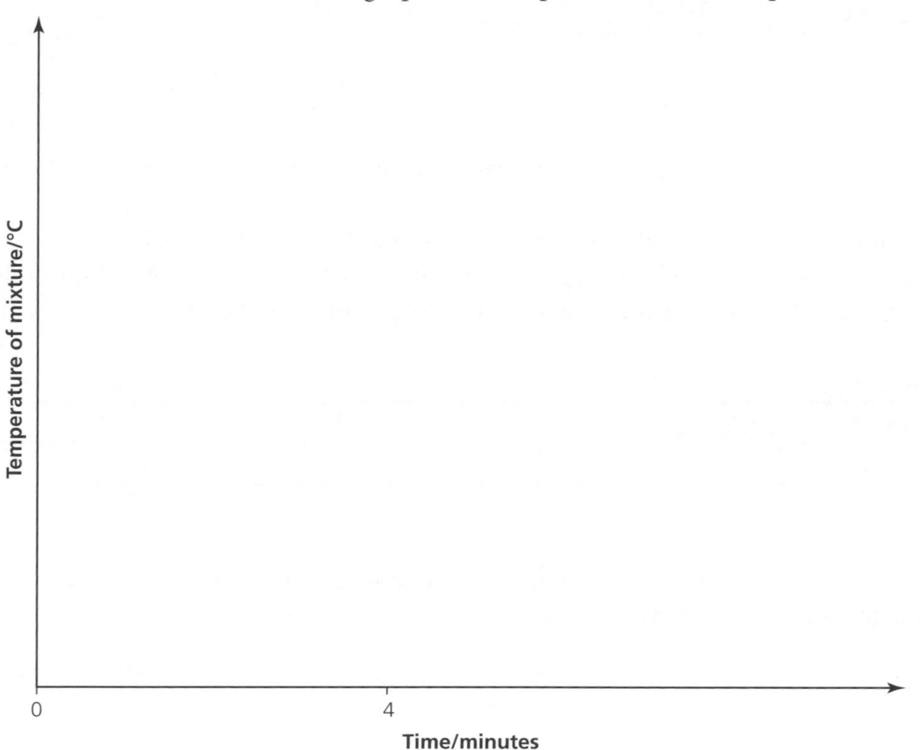

i Use ΔH_1 from **d** and ΔH_2 from **h** to determine the enthalpy change of reaction, ΔH_r, for the decomposition of $NaHCO_3(s)$.

$$2NaHCO_3(s) \rightarrow Na_2CO_3(s) + H_2O(l) + CO_2(g)$$

An energy cycle has been drawn for you.

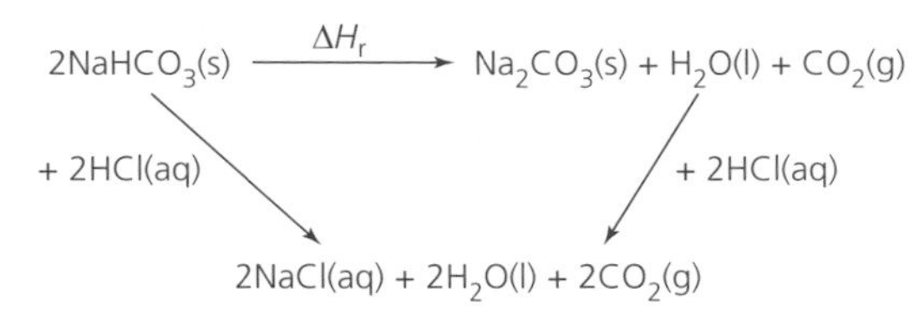

If you were unable to calculate ΔH_1 in **d**, assume $\Delta H_1 = -26.7\,kJ\,mol^{-1}$. This is **not** the correct value of ΔH_1.

ΔH_r = ... $kJ\,mol^{-1}$ [2]

[Total: 18]

Cambridge International AS & A Level Chemistry 9701 Paper 52 Q1 March 2017

11 A student was given a sample of an unknown Group 2 chloride. The student dissolved 3.172 g of the unknown Group 2 chloride in distilled water in a beaker and added an excess of aqueous silver nitrate, $AgNO_3(aq)$, to the beaker.

A white precipitate of silver chloride formed.

a Write the ionic equation, including state symbols, for the reaction occurring. [1]

....

....

b To separate the filtrate from the residue, filtration can be carried out using gravity or by using reduced pressure.

The student decided to filter the mixture under reduced pressure.

i Complete the labelled diagram to suggest how the student could filter the mixture under reduced pressure. [2]

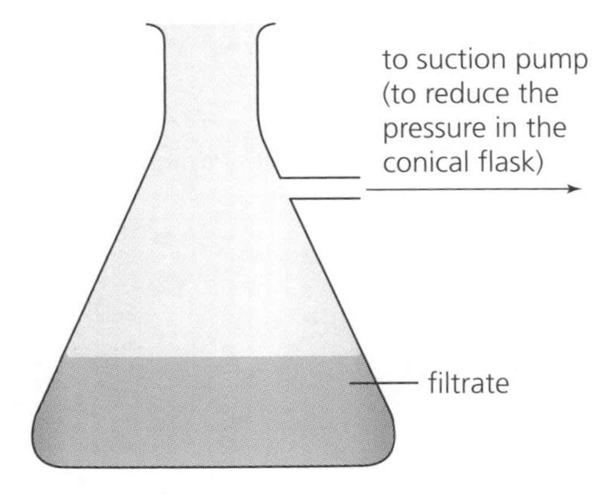

ii Suggest **one** major advantage of filtering the mixture under reduced pressure compared with filtering using gravity. [1]

....

....

c The student rinsed the residue, transferred it to a crucible and placed it in a warm oven to dry it.

i What should the student do to ensure that the drying process is complete? [1]

....

....

ii The student recorded the masses shown in the table.

Complete the table to calculate the mass of dry silver chloride formed. Use this value to determine the number of moles of silver chloride formed.

[A_r: Cl, 35.5; Ag, 107.9]

mass of crucible + dry silver chloride/g	24.898
mass of empty crucible/g	19.162
mass of dry silver chloride/g	

moles of silver chloride formed = ... mol [1]

iii Use your answer to **ii** to calculate the mass of **one** mole of the Group 2 chloride and hence identify the Group 2 metal present in the chloride. If you were unable to calculate an answer in **ii**, assume that 0.0304 mol of silver chloride formed. This is **not** the correct value.

[A_r: Be, 9.0; Mg, 24.3; Ca, 40.1; Sr, 87.6; Ba, 137.3]

mass of **one** mole of the Group 2 chloride = ... g

identity of the Group 2 metal = ... [3]

iv State and explain how the number of moles of silver chloride formed in **ii** would change if the student used tap water instead of distilled water to dissolve the Group 2 chloride. [1]

....

....

....

....

[Total: 10]

Cambridge International AS & A Level Chemistry 9701 Paper 52 Q2 March 2019

Qualitative analysis notes

Reactions of aqueous cations

Cation	Reaction with NaOH(aq)	Reaction with NH_3(aq)
aluminium, Al^{3+}(aq)	white ppt. soluble in excess	white ppt. insoluble in excess
ammonium, NH_4^+(aq)	no ppt. ammonia produced on heating	–
barium, Ba^{2+}(aq)	faint white ppt. is observed unless [Ba^{2+}(aq)] is very low	no ppt.
calcium, Ca^{2+}(aq)	white ppt. unless [Ca^{2+}(aq)] is very low	no ppt.
chromium(III), Cr^{3+}(aq)	grey-green ppt. soluble in excess giving dark green solution	grey-green ppt. insoluble in excess
copper(II), Cu^{2+}(aq)	pale blue ppt. insoluble in excess	blue ppt. soluble in excess giving dark blue solution
iron(II), Fe^{2+}(aq)	green ppt. turning brown on contact with air insoluble in excess	green ppt. turning brown on contact with air insoluble in excess
iron(III), Fe^{3+}(aq)	red-brown ppt. insoluble in excess	red-brown ppt. insoluble in excess
magnesium, Mg^{2+}(aq)	white ppt. insoluble in excess	white ppt. insoluble in excess
manganese(II), Mn^{2+}(aq)	off-white ppt. rapidly turning brown on contact with air insoluble in excess	off-white ppt. rapidly turning brown on contact with air insoluble in excess
zinc, Zn^{2+}(aq)	white ppt. soluble in excess	white ppt. soluble in excess

Table 8.1

Reactions of anions

Anion	Reaction
carbonate, CO_3^{2-}	CO_2 liberated by dilute acids
chloride, Cl^-(aq)	gives white ppt. with Ag^+(aq) (soluble in NH_3(aq))
bromide, Br^-(aq)	gives cream/off-white ppt. with Ag^+(aq) (partially soluble in NH_3(aq))
iodide, I^-(aq)	gives pale yellow ppt. with Ag^+(aq) (insoluble in NH_3(aq))
nitrate, NO_3^-(aq)	NH_3 liberated on heating with OH^-(aq) and Al foil
nitrite, NO_2^-(aq)	NH_3 liberated on heating with OH^-(aq) and Al foil; decolourises acidified aqueous $KMnO_4$
sulfate, SO_4^{2-}(aq)	gives white ppt. with Ba^{2+}(aq) (insoluble in excess dilute strong acids); gives white ppt. with high [Ca^{2+}(aq)]
sulfite, SO_3^{2-}(aq)	gives white ppt. with Ba^{2+}(aq) (soluble in excess dilute strong acids); decolourises acidified aqueous $KMnO_4$
thiosulfate, $S_2O_3^{2-}$(aq)	gives white ppt. slowly with H^+

Table 8.2

Tests for gases

Gas	Test and test result
ammonia, NH_3	turns damp red litmus paper blue
carbon dioxide, CO_2	gives a white ppt. with limewater
hydrogen, H_2	'pops' with a lighted splint
oxygen, O_2	relights a glowing splint

Table 8.3

Tests for elements

Element	Test and test result
iodine, I_2	gives blue-black colour on addition of starch solution

Table 8.4

The periodic table of elements

Group 1	2	3	4	5	6	7	8	9	10	11	12	13	14	15	16	17	18
							1 **H** hydrogen 1.0										2 **He** helium 4.0
3 **Li** lithium 6.9	4 **Be** beryllium 9.0			**Key** atomic number **atomic symbol** name relative atomic mass								5 **B** boron 10.8	6 **C** carbon 12.0	7 **N** nitrogen 14.0	8 **O** oxygen 16.0	9 **F** fluorine 19.0	10 **Ne** neon 20.2
11 **Na** sodium 23.0	12 **Mg** magnesium 24.3											13 **Al** aluminium 27.0	14 **Si** silicon 28.1	15 **P** phosphorus 31.0	16 **S** sulfur 32.1	17 **Cl** chlorine 35.5	18 **Ar** argon 39.9
19 **K** potassium 39.1	20 **Ca** calcium 40.1	21 **Sc** scandium 45.0	22 **Ti** titanium 47.9	23 **V** vanadium 50.9	24 **Cr** chromium 52.0	25 **Mn** manganese 54.9	26 **Fe** iron 55.8	27 **Co** cobalt 58.9	28 **Ni** nickel 58.7	29 **Cu** copper 63.5	30 **Zn** zinc 65.4	31 **Ga** gallium 69.7	32 **Ge** germanium 72.6	33 **As** arsenic 74.9	34 **Se** selenium 79.0	35 **Br** bromine 79.9	36 **Kr** krypton 83.8
37 **Rb** rubidium 85.5	38 **Sr** strontium 87.6	39 **Y** yttrium 88.9	40 **Zr** zirconium 91.2	41 **Nb** niobium 92.9	42 **Mo** molybdenum 95.9	43 **Tc** technetium –	44 **Ru** ruthenium 101.1	45 **Rh** rhodium 102.9	46 **Pd** palladium 106.4	47 **Ag** silver 107.9	48 **Cd** cadmium 112.4	49 **In** indium 114.8	50 **Sn** tin 118.7	51 **Sb** antimony 121.8	52 **Te** tellurium 127.6	53 **I** iodine 126.9	54 **Xe** xenon 131.3
55 **Cs** caesium 132.9	56 **Ba** barium 137.3	57–71 lanthanoids	72 **Hf** hafnium 178.5	73 **Ta** tantalum 180.9	74 **W** tungsten 183.8	75 **Re** rhenium 186.2	76 **Os** osmium 190.2	77 **Ir** iridium 192.2	78 **Pt** platinum 195.1	79 **Au** gold 197.0	80 **Hg** mercury 200.6	81 **Tl** thallium 204.4	82 **Pb** lead 207.2	83 **Bi** bismuth 209.0	84 **Po** polonium –	85 **At** astatine –	86 **Rn** radon –
87 **Fr** francium –	88 **Ra** radium –	89–103 actinoids	104 **Rf** rutherfordium –	105 **Db** dubnium –	106 **Sg** seaborgium –	107 **Bh** bohrium –	108 **Hs** hassium –	109 **Mt** meitnerium –	110 **Ds** darmstadtium –	111 **Rg** roentgenium –	112 **Cn** copernicium –	113 **Nh** nihonium –	114 **Fl** flerovium –	115 **Mc** moscovium –	116 **Lv** livermorium –	117 **Ts** tennessine –	118 **Og** oganesson –

Lanthanoids	57 **La** lanthanum 138.9	58 **Ce** cerium 140.1	59 **Pr** praseodymium 140.9	60 **Nd** neodymium 144.4	61 **Pm** promethium –	62 **Sm** samarium 150.4	63 **Eu** europium 152.0	64 **Gd** gadolinium 157.3	65 **Tb** terbium 158.9	66 **Dy** dysprosium 162.5	67 **Ho** holmium 164.9	68 **Er** erbium 167.3	69 **Tm** thulium 168.9	70 **Yb** ytterbium 173.1	71 **Lu** lutetium 175.0
Actinoids	89 **Ac** actinium –	90 **Th** thorium 232.0	91 **Pa** protactinium 231.0	92 **U** uranium 238.0	93 **Np** neptunium –	94 **Pu** plutonium –	95 **Am** americium –	96 **Cm** curium –	97 **Bk** berkelium –	98 **Cf** californium –	99 **Es** einsteinium –	100 **Fm** fermium –	101 **Md** mendelevium –	102 **No** nobelium –	103 **Lr** lawrencium –

Reinforce learning and deepen understanding of the key practical skills required by the Cambridge International AS & A Level Chemistry (9701) syllabus; an ideal course companion or homework book for use when carrying out and analysing practical work throughout the course.

- Support students' learning and provide guidance on practical skills with extra practice questions and activities
- Offer advice on key experimental methods with worked examples and diagrams
- Keep track of students' work with ready-to-go write-in exercises which once completed can also be used to recap learning for revision
- Answers can be found at **hoddereducation.com/cambridgeextras**

Use with *Cambridge International AS & A Level Chemistry Student's Book Second Edition*

9781510480230

For over 25 years we have been trusted by Cambridge schools around the world to provide quality support for teaching and learning. For this reason we have been selected by Cambridge Assessment International Education as an official publisher of endorsed material for their syllabuses.

Working for over 25 YEARS WITH Cambridge Assessment International Education

This resource is endorsed by Cambridge Assessment International Education

- ✓ Provides learner support for the syllabus for examination from 2022
- ✓ Has passed Cambridge International's rigorous quality-assurance process
- ✓ Developed by subject experts
- ✓ For Cambridge schools worldwide

HODDER EDUCATION
www.hoddereducation.com

ISBN 978-1-5104-8285-2
9 781510 482852